D0276984

GUNPOWDER & GEOMETRY

GUNPOWDER & GEOMETRY

BENJAMIN WARDHAUGH

The Life of Charles Hutton:
Pit Boy, Mathematician
and Scientific Rebel

WILLIAM
COLLINS

William Collins
An imprint of HarperCollins*Publishers*
1 London Bridge Street
London
SE1 9GF

www.WilliamCollinsBooks.com

First published in Great Britain by William Collins in 2019

1

A catalogue record for this book is available from the British Library

ISBN 978-0-00-829995-8 (hardback)
ISBN 978-0-00-829996-5 (trade paperback)

Typeset by Palimpsest Book Production Ltd, Falkirk, Stirlingshire

Printed and bound in Great Britain by CPI Group (UK) Ltd, Croydon CR0 4YY

In memory of Jackie Stedall

Contents

1

Out of the Pit

August 1755. Newcastle, on the north bank of the Tyne. In the fields, men and women are getting the harvest in. Sunlight, or rain. Scudding clouds and backbreaking labour.

Three hundred feet underground, young Charles Hutton is at the coalface. Cramped, choked with dust, wielding a five-pound pick by candlelight. Eighteen years old, he has been down the pits on and off for more than a decade, and now it looks like a life sentence. No unusual story, although Charles is a clever lad – gifted at maths and languages – and for a time he hoped for a different life.

Many hoped. Charles Hutton, astonishingly, would actually live the life he dreamed of. Twenty years later you would have found him in Slaughter's coffee house in London, eating oysters with the president of the Royal Society. By the time he died, in 1823, he was a fellow of scientific academies in four countries, while the Lord Chancellor of England counted himself fortunate to have known him. Hard work, talent, and no small share of luck would take Charles Hutton out of the pit to international fame, wealth, admiration and happiness. The pit boy turned professor would become one of the most revered British scientists of his day.

This book is his incredible story.

∞

Newcastle upon Tyne occupies a fine site for a town: a long south-facing slope, down towards the river. The Great North Road runs through it: 90 miles north to Edinburgh, 250 south to London. And the river joins the town to the world. As the old song goes,

> *Tyne river, running rough or smooth,*
> *Brings bread to me and mine;*
> *Of all the rivers north or south*
> *There's none like coaly Tyne.*

By the time coaly Tyne passes under the bridge at Newcastle it has seen the Pennines and the wild country of Northumbria: land fought over by the Romans and the Picts, the English and the Scots. If the water had once been dyed with their blood, by now it was dyed with coal.

The town was spacious and populous. Only three English towns were bigger; but still Newcastle was no sprawl early in the eighteenth century. Just five main streets within the old town walls, and open country beyond them to east and west.

∞

We know next to nothing about Charles Hutton's parents, Eleanor and Henry. It's probable they moved to prosperous, growing Newcastle from elsewhere: possibly from Westmorland, on the other side of the Pennines.

Eleanor and Henry seem not to have prospered in any tremendous way in Newcastle, though neither were they at the bottom of the heap. Henry was a 'viewer' in the collieries: something less, that is, than a land steward or estate agent, and something more than a manual labourer. Viewers were literate, numerate men who kept the colliery's records and compared notes with their colleagues at neighbouring mines. They measured and calculated rates of production, kept an eye on how fast the pit was filling with water and how well the pumps were coping. They allocated labour, planned, inspected. Some stayed up at night to catch coal thieves.

Viewers were expected to be down the pits daily, but also to see the estate agent daily. They faced two very different worlds: the wealthy owners and the men who hewed the black diamonds out of the rock beneath their feet. They had responsibility and power; some abused it, keeping owners ignorant about the mine work. Many used their position to command high fees. The coal industry was expanding, and new shafts and new mines depended on the advice of experienced colliery viewers. High-end viewers had their own assistants and apprentices: minor deities in their own field.

One source says Henry Hutton was a land steward for Lord Ravenscroft, a step further up still. We don't know if it was true, and it may just be a garbled reflection of the fact that he oversaw a mine or mines on that nobleman's land in Northumberland. Either way, he was a man who had a good deal more than the very minimum.

By 1737 Eleanor and Henry lived with their children – three or four boys, perhaps more – in a thatched cottage on the northern edge of Newcastle. Nineteenth-century historians would be rather sniffy about this 'low' dwelling (low, probably, both literally and metaphorically); but compared with the cottages of the miners themselves it was luxury. Thatch, indeed, was a luxury, as was the ability to live more than a stone's throw from the pit mouth.

Just as Henry Hutton worked at – was – the interface between landowner and manual workers, his house stood on the edge in more than one sense. Sidegate, their street, was one of the first to burst out of the still intact medieval town walls, and while open land surrounded it on both sides, a stroll down the hill took you back to the bustle of Newcastle. Away to the north it led across the Great Northern coalfield, engine of England's prosperity.

The 'low' cottage of Hutton's birth.

Charles, Henry's youngest son, came into the world on 14 August 1737. A fortnight or so later his parents took that stroll downhill to the parish church of St Andrew's, just inside the town wall – stubby tower, Norman interior, bright new set of bells – and he was baptised.

For a few years, all was well. There were games with other children in the street; there were fights. Ladies liked little Charles; they thought him uncommonly docile and good-natured.

The Huttons sent some, perhaps all, of their children to school. On the walk from their cottage to St Andrew's they passed Gallowgate (actual hangings were a rarity by Hutton's time: one every few years, if that). And on the corner of Gallowgate a house stuck out awkwardly into the street. In it an old Scots woman kept what the usage of the day optimistically called a school. That is, she took in a few of the local children and endeavoured to show them how to read, with the aid of a Bible. Charles Hutton remembered her as no great scholar; when she came to a word she couldn't read, she would tell the children to skip it because 'it was Latin'. But he did learn to read.

∞

In the summer of Charles's sixth year his father died. Swiftly his mother remarried; with several children to keep, the youngest of them five years old, there was perhaps little real choice. Their new father, Francis Frame, occupied a lower station than Henry Hutton. He was an 'overman' in one of the collieries: a labourer, but one who was also employed to superintend the others. Not necessarily literate, not necessarily numerate: there was a system of marks and tally sticks to avoid the need for that. His pay might have been half that of a viewer, and he basically worked underground. And he had to live near the pit. Shifts might start in the small hours, even at midnight, and you couldn't be walking a mile or two in the dark just to get to the colliery. The family left the house in Newcastle, and moved north into the coalfield itself. And coal changed from a distant background, a what-my-dad-does, to an immediate daily reality for Charles Hutton.

Hundreds of thousands of tons of it went down the river each year, and the market was growing all the time. Over the course of the eighteenth century the industrial revolution and the steam engine would increase Britain's hunger for coal to stupendous levels. As the coal trade grew it expanded away from the river, north and south; new collieries were opened and new wagon routes built to serve them. Landowners gambled huge sums on the chance of reaching a profitable seam, but they also told themselves – and believed – that they were doing a public service: providing jobs, providing coal.

The so-called Grand Alliance of colliery owners was sinking new pits on north Tyneside, where coal to last hundreds of years lay undug. The villages there – Jesmond, Heaton, Long Benton – would now be the Hutton family's world. This was the true coal country. The landscape of the Great Northern coalfield was all pits and pit villages; the hills and the fields, the river and the tearing clouds formed little more than a scenic backdrop. Streams were every-where, making steep little valleys and shaping where you could build, where you could walk. And where you could dig. The wet got into the pits, and a mine on north Tyneside could raise a dozen or more times as much water as it did coal. Horses plodded around horse-gins to wind the ropes that brought it all up, and unwind them again to send the men down.

The members of the Grand Alliance were technophiles, and they installed the new atmospheric engines designed by Thomas Newcomen – forerunners of the steam engine – at their pitheads to do some of the winding work. The machines were cheaper to run than horses, but they were loud, and they added to the noise, the smoke and the dust. It's hard to get a sense of it today, now that the pits are closed and the collieries built over with smart modern houses. On a clear day you can see down from Long Benton, where the Huttons worked, to the spires of Newcastle's churches, and beyond to the rolling country south of the river. There are still some open fields nearby – playing fields, now – and the long line of the old wagonway, down towards the river, remains as a peaceful footpath; birds sing in the hedges.

In the 1740s it was a very different place. The wagonway had wooden rails, and the Long Benton colliery and its neighbours ran an endless procession of horse-drawn wagons on them, taking the coal down to the staithes at the river's edge. The spires of Newcastle may have been visible through the clouds of coal dust and smoke, but they represented a distant, unattainable world.

Nearly all collieries provided housing for the underground workers, and the Huttons presumably lived in such a dwelling. Long gone now, the miners' cottages were packed in tightly, in rows as close to the pit mouth as could be managed. They were drab and uniform; two rooms was usual, with an extra chamber in the roof. Quite a few had gardens for vegetables, or even a pig or cow, or some hens. The tradition on the Great Northern coalfield – at least later – was for growing prize leeks.

The pit villages were isolated physically and socially. It was an unusual environment, supporting an unusual kind of work. The miners made a close-knit world, with massive revelry at weddings and christenings (and any other excuse): roaring bagpipes and roaring men; hilarious, vulgar, open-hearted.

Middle-class visitors typically found the miners and their families shocking, barbarous, uncivilised. Translated, that probably means miners' pubs were noisy, their homes not absolutely spotless, and their knowledge of scripture of a slightly less than prizewinning standard. The miners' world may have been rough and simple; squalid it was not.

They were proud of their appearance. Like sailors, you could tell the miners anywhere by their clothes: checked shirts, jackets and trousers with a red tie and grey knitted stockings. Long hair in a pigtail. And of course the pall of coal dust, almost impossible to wash out completely. Rings of it circled their eyes, and it made the cuts and abrasions of underground work heal into distinctive blue scars.

∞

Charles Hutton would remember his stepfather Francis Frame as a kind man. But he was not Lord Ravenscroft's land steward, nor anything like it, and there was no avoiding the fact that the family's expectations had fallen. There had probably never been any question about what work the boys would go into, but if there had been hopes of climbing the ladder to become viewers and overmen they probably now hoped for nothing more than the status of plain coal hewers.

And by the age of six, or certainly by the age of eight, little Charles was old enough to start work. One report says he spent some time operating a trapdoor in the pits. For a miner's lad this was typical entry-level work, done by the youngest boys. Different areas of the pit were isolated from each other by trapdoors, usually kept shut to make sure the air flowed where it was supposed to and reduce the risk of gas build-up and explosion. Each trap had a boy to pull the cord that opened it, and shut it again after men or coals had passed. Boring but crucial work; mines blew to smithereens if a trapper lad fell asleep on the job and let the bad air reach the candles. Solitude, silence and darkness worse than any prison. You learned not to fear the dark.

You learned much else, too; starting in the pits was not so much a training as total immersion in a culture. The all-male environment had its traditional ways: ancient facetiousness and long-lived jokes; bravado in the face of shared danger. Men died in the pits, often: candles in the gassy dark (there were no safety lamps as yet); shaft collapses; suffocation. Firedamp and chokedamp, the miners called the bad airs, and they were especially common in the northern pits. By the 1750s coal owners were asking the newspapers not to print reports of pit explosions; they were bad for morale, bad for trade.

So you learned the feel of the mines and the smell of the different airs, good and bad. You gained the true pitman's instinctive sense of danger. You also learned to cling to the rope that pulled you up and let you down the shaft. Shafts could be a few hundred feet deep, and there were no cages to ride in. You just clung on, with your leg through a loop of rope. If your hands grew tired, you died. If the horses pulling the rope were startled and bolted, you died. Eventually that would happen to Francis Frame, in an accident at Long Benton colliery after Charles had left. Over the course of a life's work in the pits your chance of dying in an accident was probably as high as one in two.

By the time the exploitation of children in the pits became the subject of official inquiry – in the 1840s – it had attained the proportions of a national disgrace. Boys were working twelve- or even eighteen-hour shifts that started at midnight; they were seeing the sun on only one day a week and they were getting rickets, bronchitis, emphysema. They were prostrated by exhaustion and they were not growing properly.

There's no real reason to suppose conditions were better in the eighteenth century, but Charles Hutton was spared the very worst of it. There were schools in some of the local villages. Some pit villages had them, too, with practice varying seemingly from pit to pit: near-universal illiteracy at one; school provision built into the miners' contracts at another. Hutton attended schools. He was bright, and neighbours were already saying the lad could go far, urging his parents to keep him at school. Kept in school he was: the hope of his family, even while his brothers went down the pits.

It must have made for some friction at home and in his village. The few pennies per day he would have earned on the trapdoors might not have been a real sacrifice for his family. But the fact that others were being prepared for a life in the coal mines and he wasn't must have made for a difference, and not perhaps a pleasant one.

About this time Hutton injured his arm. The story he would tell many years later involved variously an accident or a quarrel with some children in the street. Nothing more than ordinary horseplay, perhaps, but by the time he confessed it to his parents the bone wouldn't set properly, and it left him with a lasting weakness in his right elbow. Another reason to train his mind rather than his hands, at least for now.

A schoolteacher named Robson taught him to write, at a school a short step across the hill in the village of Delaval. Hutton may well have studied from a new, locally produced grammar book. Written and printed in Newcastle, Anne Fisher's *New Grammar* promised to teach spelling, syntax, pronunciation and even etymology through its carefully graded series of exercises. Unlike other grammar teachers she kept it practical, and there was time set aside in her programme for taking dictation from a newspaper read aloud (the London *Spectator* was her preference) as well as spot-the-mistake games. Like every textbook author, she hooked her learners with aspirational promises. By the time you were done, she said, you'd be able to write as correctly as if for the press, engage in polite and useful conversation, and compose a properly styled letter to any person of quality. London and its cosmopolitan values were never far from her thoughts. Yet she had her other foot firmly in the North, and her correct pronunciation was a distinctly northern one. Say the following words, she advised, as though the *o* was a *u*: *Compasses, Conjure, London.*

As well as his schooling, young Hutton was indulged with pennies for books of stories, and – perhaps more precious – time in which to read them. He was fond of the so-called 'border ballads', the traditional songs of north Tyneside and the Scottish Borders: True Tom and his visit to Elfland, Tam Lin and his rescue from the fairies. By his early teens one of his lifelong habits was already in place: book collecting.

∞

The routine of these years was disrupted more than once by events from outside the North-East. In September 1745, when Hutton was eight, the southward march of the Bonnie Prince and his army sparked panic in Newcastle. Some citizens hastily signed a pledge of loyalty to King George. Others spent their time walling up the town gates and mounting cannon to repel the Jacobite horde. Some fled from the northern villages to the dubious safety of the town. Others fled further south with all they could carry.

The events developed as farce rather than tragedy as far as Newcastle was concerned. Charles Stuart and his army came nowhere near; they took a western, not an eastern route down through England. The gates were unbricked, the cannon dismounted, and the King's soldiers moved on. People came back to their homes and their work, some of them presumably feeling rather shamefaced.

One eyewitness to the '45 in Newcastle was a visitor whose presence would ultimately wreak rather more upheaval, both for Newcastle and for Charles Hutton. John Wesley first visited the town in 1742. He was one of the middle-class commentators who was shocked by the drunkenness, cursing and Sabbath-breaking he found there, and he considered the field ripe for his mission.

He preached in the fields, and he preached in the churches, including the Huttons' old parish church of St Andrews (where he found the congregation notably well-behaved). He visited some of the pit villages, and preached there too. Over time, Newcastle became John Wesley's northern headquarters, the third point of a triangle whose base was London and Bristol, and he would visit again and again over the years.

The founder of Methodism was a small neat man in a gown and bands; he had been an Oxford tutor and he was good at calm, reasonable argument. But he had also been in the mission fields in North America, and he knew how to reach his hearers' hearts, with an explosive combination of plain language and restrained rhetoric.

The results were extraordinary, with people crying out from a piercing sense of their sins or falling down in dread of the wrath of God. The dramatic personal changes, or some of them, lasted long after Wesley himself was gone. Charles Hutton was deeply impressed – he started to think of himself as a Methodist, and to call himself one.

It wasn't about leaving the Church of England: that came much later for Wesley's supporters. But it was about reinventing yourself and your relationship with God: about getting a new sense of what a life and a self could be. It was Christianity made both primitive and experimental, with doctrine founded on scripture, experienced and confirmed emotionally, and integrated into your personal habits, into who you were both internally and visibly.

Several of the anecdotes we have about Charles Hutton in his youth are concerned with his piety. He threw away his collection of profane stories. He built a cabin in the woods where he could pray on the way to school. He read devotional tracts. Time would eventually lower the temperature of Hutton's enthusiasm, but he would remain a follower of Methodism – later shifting towards Unitarianism – until his thirties.

Some of the practical characteristics he gained at this time he would take to his grave, and they laid the foundation of much he achieved as an adult. 'Never be unemployed for a moment,' wrote Wesley; 'never be triflingly employed.' Charles Hutton would retain into old age a reputation for the good ordering of his time and his thoughts. Hard-working, self-disciplined, cheerful yet grave, and gifted at organising both ideas and people, he could have been a model for such works as Wesley's *Character of a Methodist*. But perhaps the most important lesson Charles Hutton took from Wesley was that you could reinvent who you were, remake your mental world and your character. You could forge a destiny of your own choosing, both in the next world and – perhaps – in this.

By his early teens Hutton was living with his family in the village of Heaton, and attending school half a mile away at Jesmond. His teacher, Jonathan Ivison, was a would-be clergyman: a university graduate marking time while he waited for his first benefice.

Ivison represented another link to a wider world for Charles Hutton. He was no spectacular example of worldly success, but he illustrated the very different kind of life that could be reached through education. And his teaching provided direct access to parts of that education. He taught Hutton some Latin and some mathematics, and Hutton discovered talents for both. These were whole new worlds to explore: ancient literature or the heady abstractions of algebra and geometry could easily occupy a person for a lifetime. Meanwhile Hutton's ability at accounting drew some attention, and presumably provided evidence to his family that keeping him in school remained a reasonable choice. And he was also learning practical geometry and surveying: food for the mind, but skills that could one day be of use in the world of the collieries, too.

His continued success at school reportedly made Hutton a favourite with those who taught him. He seems to have become quite close to Ivison. Ambitious, always keen to be at the top of everything in hand, he was envied by other students.

But time was running out. Hutton couldn't remain the hope of his family for ever, and unless an unexpected opportunity opened up the only route his family or his world could offer him was a career in the coal pits, perhaps working up to overman or viewer if he was very lucky. At fourteen he passed the age when schooling for boys of his class ended and work or apprenticeship began; and it wasn't clear what was going to happen next.

Indeed, it's not clear what did happen next. Jonathan Ivison gained his benefice in the autumn of 1751, when Hutton was fourteen, and after hasty ordination he was licensed as curate at Whitburn, near the coast in County Durham, a couple of days later. Whitburn was nine miles from the schoolroom at Jesmond – a dozen by road – and commuting between the two was scarcely feasible. Schools were transient things, and there was no particular reason this one would stay open once its single teacher was gone.

Yet Ivison's new salary was just twenty-five pounds a year: barely enough to live on. He badly needed more money. One report says Hutton acted as Ivison's assistant. It's possible he acted as his substitute, keeping the school open while Ivison performed his parish duties half a day away. There was nothing awfully unusual about the head of the class helping to teach the younger children, and nobody ever denied that Hutton was the head of this and every other class he ever entered.

Whatever the truth of the matter, Hutton was spared a return to the pits for a few more years. There's no hint in any of our sources that he ever did the kind of work to which teenagers normally graduated in the mines: 'putting', that is, dragging or pushing the heavy wicker baskets full of coal from the face to the shaft.

But his reprieve was not to last indefinitely. The next firm fact we have about him is a pay bill from the Long Benton colliery for September 1755, when he was eighteen. He was working as a coal hewer in the 'Rose' Pit, under his stepfather Francis Frame as overman.

∞

Put one leg through a loop in the rope, and hold it hard for the breathless minutes of descent: three hundred feet down. At the shaft bottom the darkness is total, outside the reach of a few brief candles. Long, dreary galleries lead off in every direction, and it's oppressively warm. Walls of coal; wet under foot. Fossils in the roof to remind you of the weight of rock, the inexorable crushing force ever-present above your head. Constant noise: rushing water and the distant thud of the engines; corves of coal being dragged past by teams of exhausted, shattered boys. Grimy, weary men everywhere, bustling in the gloom.

Strip to the 'buff': short breeches, low shoes, cotton skull-cap (or even less). Take your wedge, hammer and five-pound pick. Kneel, sit, stoop to get at the coalface. Or just lie down. Coal seams could be as little as two foot six high, and Hutton was a

opposite: At the coalface.

tall young man. Channels of sweat in the coal dust on your face.

First, make vertical cuts from the top of the seam to the bottom, dividing the coalface into what the men called 'juds'. Then, undercut the juds one by one: skilled and, by some accounts, athletic work. Next hammer in wedges at the top of the jud, and bring down the coal section by section. It could take an hour or two, and then, of course, you did it again. And again, all day. A hewer could bring down four tons or more in a shift. Putters bundled it away in the wicker baskets, and up it went to the daylight.

It was hard, it was dirty, it was dangerous. But coal hewers were actually the elite of the pit; indeed, it's surprising Hutton was allowed to do the work as young as eighteen, and without working his way up through gruelling years at the back, and then the front, of a corf of coal. Francis Frame perhaps used his own position in the pit to make it happen. Hewers were relatively well paid: quite possibly better paid than manual workers in other fields like agriculture. And they worked shorter hours than the trapper boys and

the putters: seven or eight hours of the day, starting in the very early morning.

> The stars are twinkling in the sky,
>   As to the pit I go;
> I think not of the sheen on high,
>   But of the gloom below.
>
> Not rest or peace, but toil and strife,
>   Do there the soul enthral;
> And turn the precious cup of life
>   Into a cup of gall.

That was Joseph Skipsey, the 'Pitman Poet', in the 1860s: but the words could have applied to Hutton, or to many and many another.

Charles Hutton had known, briefly, the life of a country school-teacher, and had felt the stirrings of a talent for both languages and mathematics. A world wider than the pit had come into view more than once during his short life, in the person of John Wesley and in that of Jonathan Ivison: university graduates both, and holders in their different ways of promises about what life could be.

In a pamphlet for children, printed in Newcastle, Wesley wrote that you shouldn't ever wish for more than you have. Men are fallen, grace is a gift, and if we had what we deserve we'd all be in hell. But however docile Hutton was, and however resigned to the will of Providence, it's hard to imagine he didn't wish.

I think Charles Hutton's return to the coal pit must have represented desperation – perhaps exasperation – on the part of his family. The hope that if he stayed in school he might somehow make a better or a different life for himself was showing no clear sign of paying off, so far as we know. Children younger than him were earning their keep; the favoured youngest son had been indulged for many years, but he would not be indulged for ever.

One source says he was 'taken from school', and there is almost the sense of a violent abduction away from his natural element to the now quite alien world of the pits.

The pit, and its cup of gall, could have been the rest of his life: for a time there must have been no reason to imagine anything else was going to happen. Two things made a difference.

One was that Hutton was bad at hewing coal. The pay bills show him bringing down less coal than his fellows, and as a result he was employed less – much less – than they were. In old age Hutton would say his injured arm was to blame for his poor performance at the coalface. Maybe it was, though we never hear that the lame arm affected him later in his life. Maybe it was simply that his schooling and his comparative lack of experience down the mines had left him without the strength, the endurance and the skills the work needed.

The other thing was that Hutton was using his time away from work to exploit his connections, to invent a way out of colliery life and make something quite different for himself. Who knows just what hustling Charles Hutton did during the months he worked down the pit, in the winter of 1755–6. But it came to the point that he knew he had it in his power to escape, and some sort of scene appears to have taken place at the colliery. He was 'laid idle' one day, whether for some fault or simply because a more able man was available, and he made a fuss about it. He told the overman that he'd soon be out of his power, that he was going to go to a more respectable life. It must have been quite a row; men who were there remembered the scene seventy years later.

But Hutton was speaking the truth, however ungraciously. Jonathan Ivison had been the curate at Whitburn for four years now, and if his Jesmond schoolroom had been limping along meanwhile, it was time to pass it on for good to someone who lived near by. Ivison knew the man for the job. Presumably he squared it with the owner of the building where the school was held. And presumably he squared it with Francis Frame. For he wanted Charles Hutton to take over.

At the age of eighteen, Hutton was legally a minor and too young, in theory, to set up on his own in any trade or profession. (He was also, perhaps crucially, too young to have signed any sort of binding contract at the colliery.) But it's hard to imagine his parents having much hesitation about an offer that would get the useless hewer away from the coalface, where he was little more than an obstacle in the way of fitter, stronger men, and would secure him – if he pulled it off – an income and even a career. One day in March 1756 Charles Hutton clung to the rope and ascended to day for the last time.

2

Teacher of Mathematics

Autumn 1758. High Heaton, near Newcastle. Silence (or almost).
Boys seated at rows of desks. Slate pencils squeaking; quills
scratching on paper. Another harvest is in the fields, and the coalface
is as hellish as ever. But Charles Hutton is above ground now, in
what was once Jonathan Ivison's schoolroom.

The young schoolmaster is seated at the front, where he can see
all the boys. One at a time comes forward, shows his work, is
praised or blamed. Can he spell the words set the class; has he
written his fair copy fairly enough? Can he read the passage
assigned? Has he worked the sum correctly and by the right method?
Does he know his catechism answers? He is given a new exercise
to do or a new instruction or definition to copy down in his book;
sent back to his desk.

Over the course of the day and the week the work cycles through
reading, writing, arithmetic, divinity. The boys spell from a textbook: probably the same one, by Ann Fisher, that Hutton learned
from himself. They read from the Bible, learn to count or do their
sums from another book. Start again on Monday.

It would be intriguing to know how Charles Hutton spoke to
his students. Firmly or gently? Leading or prodding? What was he
really like as a teacher? We shall never know.

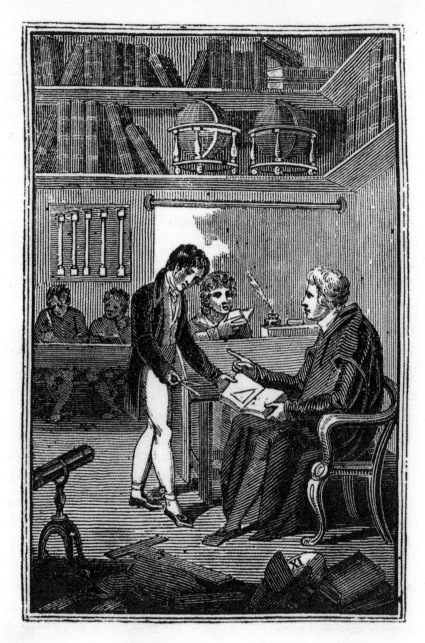

A master and his pupils.

But the books he wrote, later on, record something of the quality of thought that went into Charles Hutton's work in the classroom; they hint at a man who had his material impeccably organised, who always had something new to show the student, a new trick, a new bit of complexity to stretch the growing mind. The Georgian classroom involved lots of one-on-one teaching, most of the children working quietly while the teacher spoke with each in turn at his desk. It demanded huge agility of mind from the teacher, as each student to be seen was at a different place and needed a different sort of help.

It's easy to imagine that a teacher with a less than perfect grasp of the material, or who relied too heavily on a printed textbook, would end up with most of the students spending most of the time on repetitious exercises they only partly understood, and would quickly find he had lost the weaker ones. But for the exceptionally wide-awake teacher the system provided a maximum of flexibility and opportunity. It depended on having the material very clear in your mind, and above all minutely graded into separate steps. Hutton seems to have excelled at this. He would make a few enemies during his long life, and pick up criticism, some of it deserved. But no one ever denied he was a superb teacher.

He was talented, he was full of energy and he worked hard. The school at High Heaton ran both days and evenings, teaching local boys and adult miners, some of them Hutton's former mates from the pits. Local children flocked to him as students, to such a degree that the schoolroom proved too small. So Hutton obtained the use of a room in nearby Stote Hall, and there he and his scholars trooped daily for their lessons.

It's demolished now, the former mansion of Sir Robert Stotte; only the gateposts and gatehouse survive. But photos show an imposing house. It looked over the small valley of Jesmond Dene and you can still retrace Hutton's walk of a quarter of an hour or so from one village to the other. First a very gentle rise through what was then open farmland; then down the hill and across a steepish valley of never-farmed woodland. Sloping paths and views away through the trees. It was already a walk into a very different world from that of the colliery, and infinitely removed from the coalface in the Rose Pit.

And yet for many, schoolteaching was a despised profession, a mark of failure for a man of education. 'School', indeed, is a grand term for what was no more than a room provided by someone's goodwill and in which Hutton collected pennies for teaching young children to read and write, to add up and remember a few Bible stories. Chemist and religious controversialist Joseph Priestley was typical: 'Like most other young men of a liberal education, I had conceived a great aversion to the business of a schoolmaster, and had often said that I would have recourse to anything else for a maintenance in preference to it.' Elementary teaching in particular was widely seen as work for old women, widows, or the humblest members of society. Nor did it bring any very great material rewards for its long, demanding hours: pay only slightly above that of a labourer; no such thing as promotion; no such thing as a pension.

There was no teacher training. Giftless graduates stumbled into teaching and stumbled through it; impecunious clergymen practised it with a savage resentment that they took out physically on the children in their charge (it was the boast of Dr Parr of Harrow that he never flogged a boy twice in the same lesson). A typical contemporary observation is contemptuous of what went on in small ragged schools like Hutton's:

> Poor Reuben Dixon has the noisiest School
> Of ragged Lads, who ever bow'd to Rule;
> Low in his Price – the Men who heave our Coals
> And clean our Causeways, send him Boys in Shoals.

Indeed, teaching unruly boys just a few years younger than himself demanded more natural authority than Charles Hutton yet possessed. He would be remembered ominously as having 'kept up the most rigid order' as a teacher, and sometimes having carried 'his severity too far'. One student from these early days recalled that he 'assumed a degree of importance' in the classroom:

pomposity might have been a blunter word. For a while he affected a large academic gown and, to complete the effect, a scarlet cap. Even his best friends were embarrassed. He turned up to a parish election in this finery, and 'his friends, who would have supported him in the state of a Caterpillar, were so disgusted when they saw him transformed into a Butterfly, that they did not support him and he lost his election'.

Hutton was very young, and the phase passed. The excellence of his teaching continued. He quickly determined that the schoolroom at Stote's Hall would not be the end of his journey. With the ferocious energy and the self-discipline that would attract comment again and again, he set himself to improve still further. He read all he could: not chapbooks of romantic stories now, but the hardest and the newest mathematical books he could lay hands on. Newton's works and the works of his contemporaries and disciples: Christiaan Huygens, Roger Cotes. Descartes and his followers. Textbooks of gauging and surveying, and the works of Hutton's own contemporaries in Britain and beyond.

On top of his teaching and his reading, he went down the hill to Newcastle in the evenings and attended classes given by a Mr Hugh James, who specialised in mathematics. It was a conspicuously demanding regime, and his mother feared for his health. But Hutton could see where his future lay, and he had determined to pursue it as hard as he could, whatever the cost.

∞

In fact it was his mother that died, in March 1760. She was buried near her first husband in St Andrew's, Newcastle, on the seventeenth of that month.

Hutton was twenty-two. We don't know just how relations stood between him and his mother when she died, or even whether he had already moved out of the family home. We do know that less than three weeks later he returned to St Andrew's to be married. His bride, Isabella, was four years his senior; she had trained as a dressmaker. The marriage licence gives her maiden name as Hutton, so she may have been a relative. The scarcely decent haste hints at a family drama now lost from view: a match on which the enamoured couple were keener than were their families, perhaps. Soon there was a son, Henry, known to his parents as Harry.

They moved into rooms in Newcastle itself, just off the Flesh Market: central, bustling, but rich in shrieks and stink. And just a week after his marriage Hutton advertised a new school in the Newcastle papers.

TO BE OPENED

On Monday, April 14th, 1760, at the Head of the Flesh Market, down the Entry formerly known by the name of the Salutation Entry, Newcastle, A Writing and Mathematical School, where persons may be fully and expeditiously qualified for business, and where such as intend to go through a regular course of Arts and Sciences, may be compleatly grounded therein at large.

He promised to teach writing, arithmetic and shorthand, as well as a long list of mathematical subjects: accounts, algebra, geometry, mensuration, trigonometry, conic sections, mechanics, statics, hydrostatics, calculus. Any youths not satisfied by these would also be shown their applications to practical life: navigation, surveying, gunnery, dial-making, measuring, geography, astronomy.

It all seems premature, precocious, absurdly risky. Hutton was just four years out of the coal pits; he was quite unknown, his fees were twice those asked by his rivals and his list of subjects promised

opposite: The tree of mathematical knowledge.

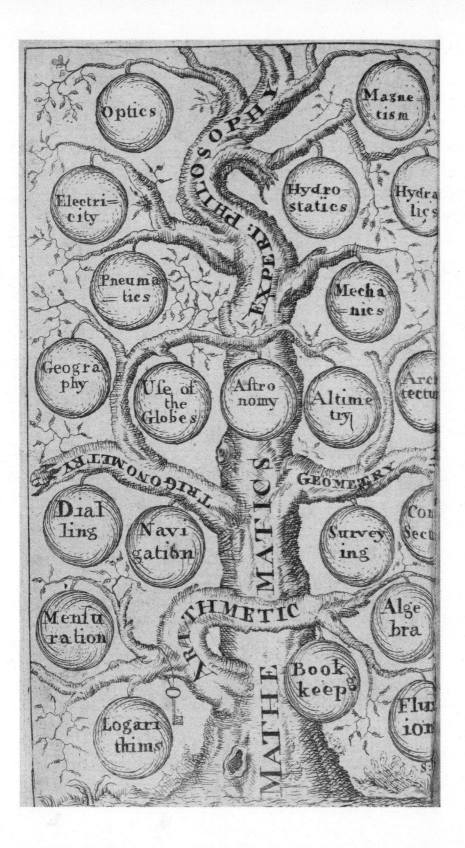

an impossible range of expertise, including any number of practical subjects in which the twenty-two-year-old had no practical experience. He was aiming very high, hoping to become a different class of teacher: a specialist in mathematical training, no longer bound to the drudgery of teaching the very young to read and write. Friends advised him to promise less, and ask for less money. He didn't listen, and the result was a struggle, the high prices keeping some away during years when Hutton's family was growing.

But there was real demand for this sort of thing: specialist mathematics teaching for boys up to fourteen or even older. In a period when a quarter of the country was unable to read, northern parents had a reputation for being keen to educate their children, for sending able, well-educated boys to London to work in counting houses and trade. Trade was increasing during the eighteenth century, and the demand for those boys multiplied; they became bookkeepers, accountants, land stewards, surveyors, navigators (think of James Cook, who would sail round the world on a Whitby coal-ship; while he was still in the coal trade he learned all the mathematics he could at Newcastle schools between voyages). Schools proliferated, and Newcastle became England's best-educated city after London: four charity schools and the prestigious Trinity House School were founded in the first half of the century, and its well-regarded grammar school flourished. A few landowners maintained schoolmasters for the benefit of their employees, and it wasn't unknown for the employees themselves to club together to fund a school, as they did at the Newcastle ironworks. Hutton had picked a growing market.

Teaching mathematics at this level meant teaching to add, subtract, multiply and divide, to find square roots and cube roots (Hutton had his own special method for this). It meant fractions both natural and decimal: how to read and write them and how to do arithmetic with them. It meant handling England's enormously complicated systems of units: grains, scruples and drams for drugs; yards, poles and furlongs for length; firkins, kilderkins and barrels for beer; and many,

many more. It meant converting between currencies, which brought similar difficulties: how many Flemish guilders will I buy with one hundred and seventy-three pounds, fourteen shillings and twopence, if one pound is worth thirty-five stuivers three and a half penning?

Above all, it meant reasoning with proportions: or rather, for the less able who were the majority, applying a series of rote-learned rules that would tell you the right answer in certain situations to do with linked ratios. If eight yards of cloth cost twenty-four shillings, what will ninety-six yards cost? In how many days will eight men finish a piece of work that five could do in twenty-four days? In a school career a child might do hundreds, if not thousands of these problems.

It was dry stuff; but Hutton made it work, and over time the numbers at his school grew. By 1764 he moved downhill to the more fashionable, though somewhat unfortunately named 'Back-Row', where he, his school and his family shared premises with a dancing master named Stewart. Hutton got involved with selling tickets for the man's public balls. Stewart's prices (half a guinea at entrance and a guinea a quarter for six days' teaching each week) would have limited his students to the well-to-do, but still the presence of a dance school in the house can only have been disruptive, both to Hutton's own teaching and to his domestic life.

We have few glimpses of that life. In one of his books Hutton, searching for a memorable image, remarked that a triangular prism 'is something like a hat box'. Indeed it is: but hats in hat boxes were not a picture that would have sprung to Hutton's mind a few years before. An admirer remarked much later that Hutton 'was soon conscious of his great abilities, and claimed that rank in society to which they entitled him'. That was a polite gloss on the fact that material success gave him the means to act, dress, and for all practical purposes *be* middle-class. He was now indisputably the butterfly, not the caterpillar.

Still, he could not altogether avoid criticism. Some remembered 'a very modest, shy man' at this period; but we also hear of him knocking a boy down in the street while he was surveying Newcastle. In this or another incident someone taxed him with being only a pit boy, and Hutton retorted that if he – the critic – had been a pit boy he would be there still. Memories of the cap and gown lingered; two different witnesses, seventy years later, independently recalled the red cap.

There were three more children: Isabella, Camilla and Eleanor (known to her family as Ellen). They were baptised in the nonconformist chapel at Hanover Square.

At the time of writing Hanover Square is a demolition site tucked away near the line of the old town walls. Trains rattle past over the nearby viaduct, and it's hard to get a sense of what was once Newcastle's only open square. The chapel there was established in 1727; from 1767 it had a school, and by 1810, after some modification, it was large enough to hold an organ and six hundred people. Its congregation included prominent local poets, newspaper proprietors and politicians.

Hutton had remained a zealous Methodist for some time after his childhood conversion: one report says he wrote sermons and preached them, though if this is true they have – sadly – not survived. His connection with Hanover Square may mean he had now left behind a movement which at this date still aimed to reform the Church of England from within, not from without.

In fact the Hanover Square chapel would later acquire a reputation for Unitarianism, and Hutton's presence there likely signals that he and his wife had come to be interested in more radical kinds of Protestant nonconformity, mixing in circles which questioned even such traditionally core doctrines as the Trinity, the atonement, and the divinity of Jesus. His private commitments are nowhere recorded, but a wider circle of Unitarians and those with radical sympathies – religious and political – would shape Hutton's professional development long after he left Newcastle.

This could have had serious practical consequences. Probate courts, marriage, schools and universities all potentially discriminated against non-Anglicans, and the ill-named Toleration Act of 1688 specifically excluded deniers of the Trinity (as well as Catholics) from its provisions. Meanwhile the Blasphemy Act of 1698 threatened them with up to three years' imprisonment and loss of civil rights. Enforcement was patchy, but the risk of penalties was real, as was that of the loss of friends. Hutton's clerical benefactor Ivison is conspicuous by his absence from Hutton's life after his move to Newcastle.

Work continued, and at a remarkable pace. Materially, Hutton was certainly prospering. His trajectory culminated for the moment with a final move to Westgate Street, one of Newcastle's wealthier residential spots. He met the recurring problem of inadequate premises by acquiring a plot of land and building his own house and school. It was quite the elegant Georgian pile, with cellars and other conveniences. The Huttons could now avail themselves of all that the prosperous, growing city had to offer.

The old town walls (they had started to come down in 1763) enclosed an area of less than 200 acres, but those acres held the north of England's capital and a good proportion of the northern counties' population. There were tall elegant buildings and wide open spaces by the several churches. By the early 1770s the town had three hundred street lamps and a well-organised night watch. If the lower town tended to be smoky from house fires, and if the riverside was dominated by busy warehouses and the bustle of shipping on the Tyne (not to mention a growing concentration of poor tenements near the Black Gate), there were open fields just a little further up the hill, some under conversion into elegant pleasure gardens.

Newcastle in 1745.

The town offered the full range of mid-Georgian amenities. Subscription concerts, both at the assembly rooms and outdoors in the summer. Visits from famous musicians en route from London to Edinburgh, who would often stop and give a performance or two to offset the costs of travel. Charles Avison, Newcastle's own resident composer and concert promoter, a man who enjoyed national fame. A literary club; theatres. Scientific lectures: Newcastle was the first provincial town to have them, with both local talents and the nationally famous visiting on tour. A popular press: for most of the century there were two weekly papers. Newcastle had local histories and local poets and balladeers. There were shops, inns, clubs, societies; fashion, food, wine.

Ultimately it all rested on coal; you could hardly forget that, if you lived anywhere near either the river or the coalfields, and Charles Hutton was never likely to forget it either. The sale of coal

was so lucrative, contemporaries reckoned, that despite all the goods it imported Newcastle made a net gain every year, and had more money per head than anywhere else in the kingdom.

Mathematically, too, Newcastle and its environs were a rich world. There was William Emerson, a nationally famous mathematician who lived in nearby Hurworth. His studied eccentricity of manner and dress (home-made linen, big floppy hats and shapeless old jackets) earned him a local reputation as a wizard. Hutton corresponded with him and they became acquainted, though unsociable Emerson and ambitious Hutton did not really hit it off. There was John Fryer, who assisted Hutton with his surveying work; he was also Hutton's teaching assistant at Westgate Street.

The school had its own separate entrance. Advertisements, briefer and more sober now, stated that there 'Youth are qualified for the Army, Navy, Counting-house'; they could also be 'compleatly instructed in the Theory and Practice of Land Surveying, with the use of the necessary Instruments'. Newcastle Grammar School took to sending its students to Hutton for specialist mathematics teaching. Not an altogether unusual arrangement – everyone knew that private academies did mathematics and science better than the grammar schools – but a gratifying endorsement of Hutton and his work.

His new-found middle-class status also meant that Hutton could work as a private tutor to the local gentry. As one biographer put it, Hutton's 'manners, as well as his talents', now 'rendered him acceptable' in this role. Robert Shaftoe was one such patron, his home at Benwell Hall one of the more impressive local mansions (the Bobby Shaftoe of the popular song was a relative). Hutton's tuition of his children impressed Shaftoe so much that he took to attending the lessons himself, revising the mathematics he had perhaps learnt at college. And he gave the young man the run of his impressive library. Newcastle had at least a couple of subscription libraries, and no shortage of booksellers, but access to a large, private book collection was a boon for Hutton, who remorselessly continued to improve himself. Over the years he added a reading knowledge of French, Italian and German to his early-acquired Latin. By 1772 he had read enough on geography to offer public lectures in the subject (to 'gentlemen and ladies') at half a guinea for the course. How many takers he found is not recorded.

Indeed, Hutton's school was becoming something of a centre for learning. During the Christmas vacation of 1766–7 Hutton taught mathematics to other schoolteachers there, and at about the same time external lecturers began to use it as a venue. Caleb Rotherham covered geography, astronomy and other scientific subjects; likewise the popular James Ferguson. Ferguson was a house guest, and gave private performances to Hutton's friends and family in the evenings, though Hutton was shocked when he discovered how little geometry the man knew. Hutton's school was in a way a forerunner for the Literary and Philosophical Society that would be founded in Newcastle thirty years later, its first paid lecturer the same Caleb Rotherham.

His arrangement with Newcastle Grammar School brought Hutton himself two of his most celebrated pupils, and certainly his most colourful. John Scott was the son of a coal merchant; Bessie Surtees a wealthy banker's daughter. Both went to Hutton for their mathematics lessons, but it was presumably not under his eye that their youthful romance blossomed. Scott went up to Cambridge in 1772, but the calls of love proved stronger and in November he came back and eloped (ladder, first-floor window) with Bessie. The scandal – or the romantic adventure, depending on your perspective – ran and ran through the nineteenth century, and Bessie Surtees merchandise is still to be had in her Newcastle home town.

3

Author

A country clergyman, anywhere in England, any time in the eighteenth century. At his desk, in his study. April. Open windows; sunlit air. An open manuscript of mathematical work. Solutions to puzzles from magazines, copied fair.

He copies this year's set of solutions fairer still, adds a couple of suggestions for problems the magazine could ask this year. Mends his pen and adds a covering letter. *Dear Sir. I enclose my mite for this year's Diary, hoping you will find it worthy of notice. Your humble servant.* Dusts the sheets, folds, seals.

A scene repeated many times – many hundreds of times – across Britain every year of the Georgian period. A few months later, in about October, the annual magazines went on sale: *The Ladies' Diary, The Gentleman's Diary, The Mathematical Repository*, and more. Some readers had the thrill of seeing their names, their mathematical work in print – a few won small prizes. Others looked in vain for their names, their work in the magazine, and concluded, humiliated, that their solutions had been wrong.

∞

If Hutton had done no more than succeed as a provincial school-teacher, his story would be a striking but not a very unusual one. The majority of mathematics teachers in Georgian Britain, indeed, were working-class lads who had made good: self-made men who had themselves attended private academies or bettered themselves by private reading. There were schools right across the United Kingdom that bore witness to their success in attracting students, providing them with high-level instruction and sending them out to work in the burgeoning literate and numerate trades.

Hutton was not satisfied with this. He wanted the wider recognition and the promise of greater rewards that publication would bring. And he approached publication through that remarkable Georgian institution, the mathematical periodical.

Now that they have disappeared, it's hard even to imagine them, but in their heyday there were a dozen or so monthly or annual magazines whose purpose, or one of whose purposes, was to print mathematical problems and readers' solutions to them. Construct a triangle given its base, one adjacent angle, and the line bisecting the opposite angle. Find a fraction with the property that, if you subtract its reciprocal, you get a square number. How many ways can you make fifteen from a pack of cards? This was not Sudoku, and it was not elementary arithmetic. The problems could be hard, using lots of algebra and geometry and sometimes even calculus. Since both problems and solutions were sent in by readers, an inevitable show-off effect meant that over time the problems tended to grow harder, the solutions more elaborate.

Despite the difficulty of the mathematics, the magazines sold plenty of copies: thousands, even tens of thousands. At any one time there were probably several hundred readers contributing their problems and solutions to them: it's hard to say because, intriguingly, anonymity was the norm. Your name appeared in print only if you specifically said it was all right to print it. The writers were teachers, practitioners, gentlemen and women enthusiasts, schoolboys. They called themselves 'philomaths', and they loved mathematics for aesthetic and moral reasons as well as because it was, for many of them, a lifeline to a wider world of culture and ideas than they would ever reach otherwise: a world in which mathematical competence was everything. The austere language of mathematics was a very good place in which the shy, the modest and the provincial might both hide and shine, while allowing working-class and female mathematicians to contribute, perhaps anonymously, without being labelled. It enabled them to interact in a controlled way with people all across the country, to display what they were good at, improve their skills, lighten their countrified boredom.

Hutton approached this printed world under the tolerably transparent anagram of Mr Tonthu. Close to home, there was a mathematics column in the weekly *Newcastle Courant*, but he didn't touch it. Instead, starting in December 1761, he sent in a string of able, elegant solutions to problems in *Martin's Magazine of the Sciences*. (Benjamin Martin was a schoolmaster, lecturer, optician, seller of mathematical instruments, author, editor and tireless self-promoter: he issued his magazine monthly from 1754 to 1763.) He also appeared as the proposer of four questions of his own.

After two years Hutton/Tonthu became more ambitious: five of his solutions were printed in the much more prestigious *Gentleman's Diary*, an annual compilation devoted to mathematical and other puzzles, and reputedly the home of the hardest of the philomaths' problems. Hutton's questions included equations to solve, geometrical constructions, and formulae for trigonometric expressions. Then he felt it was time for a fresh start under his own name, and the world heard no more of Mr Tonthu after 1763.

This time he aimed right at the centre of philomath culture: *The Ladies' Diary*. Set up in the first decade of the century to contain charming little anagrams and easy mathematical problems in verse, the *Diary* had become under a series of editors the queen of the philomath journals, its four dozen pages containing – as well as an astronomical almanac for the year – problems as many and as hard as any of them. Some of the contributors were women, or pretended to be, but it had become primarily a place not to engage in genteel discussion of mathematics but to display high-level, up-to-date skills. Indeed, as an attempt to make mathematics a subject of polite public discourse it had by mid-century failed; like philomath culture as a whole it had become another instance of mathematics' tendency to exclude and therefore to be anything but polite.

Visibility in this world was a prize worth having, and Hutton seized it. Over a decade from 1764 to 1773 he sent *The Ladies' Diary* a total of fifty-seven correct solutions, of which twenty were printed in full. He answered the prize question correctly on five occasions: prizes were determined by lot and he won a total of twenty-eight free copies of the *Diary* for his pains. He also proposed four questions for solution by others, though two of them proved to be rather too hard and no correct solutions were received except his own. Still, if you were a British philomath in this period, you most certainly got to hear about Mr Charles Hutton of Newcastle.

Hutton would later write that an advantage of the mathematical correspondence promoted by *The Ladies' Diary* and its sisters was that 'considerable additions are made to the stock of mathematical learning in general, as well as to the particular knowledge of individuals'. Behind the scenes, he was finding other ways to add to his own stock of mathematical learning. Having already attended the schools of Hugh James in Newcastle and Mr Robson at Delaval, once in Newcastle he embarked on a systematic, historically motivated programme of mathematical reading, covering the Greeks, Romans, Spaniards, French and Germans as well as British mathematical writers.

During 1763 he distilled the fruit of his reading and his teaching experience – all six years of it – into a short textbook on arithmetic. *The School-master's Guide* was published in Newcastle on 3 March 1764.

Its subject was just what Hutton had been teaching: elementary arithmetic, beginning with addition, subtraction, multiplication and division. The book continued with proportional reasoning in all its diversity and how to find square roots and cube roots. There was little more: one of the selling points of the book was its spare, uncluttered approach. Yet the careful control with which Hutton increased the complexity of his examples, and his penchant for introducing new tricks, rules or exceptions midway through what looked like routine series of examples, certainly kept things interesting. We gain a sense of what Hutton's teaching was like in person: agile, thoughtful, tremendously well organised. Whatever exercise is being done, there's always a *slightly* harder version of it just over the page.

Indeed, one of the reasons for the *Guide*'s success was the clarity with which it presented Hutton himself as a safe, sure, capable guide to the tricky territory of beginners' arithmetic. Here was a man who loved calculation, who was almost preternaturally good at it. A man for whom common sense would unproblematically tell whether an answer was reasonable or not, for whom number sense was – as a matter of course – good enough to use obvious simplifications when the numbers in a calculation suggested them. For whom long division could be done largely in your head after a bit of practice: 'when you are pretty ready in division, you may, even in the largest divisions, subtract each figure of the product as you produce it, and only write down the remainders.'

There were a few missteps in the *Guide*, indeed, when things were evidently clear to Hutton but he was unsuccessful in setting them out lucidly in words. Some of his attempts to give verbal equivalents of algebraic rules would have been scarcely comprehensible without the help of an able teacher. Some of his special tricks complicated more than they simplified: if a multiplier is itself a product, multiply by its factors separately. If it's not a product, find a nearby number, multiply that, and then correct the answer by adding or subtracting.

But ultimately the aim of all his rules, tricks and practice examples was to impart to students something of his own feel for numbers, to help them develop a number sense and be able to select the right calculatory process even in an unfamiliar situation. And in that he appears to have succeeded.

Hutton moreover took pains to come across as a humane man, one who knew that children would get things wrong, that 'calculations of the same accounts made at different times will sometimes differ', that some pupils were simply not fitted for difficult calculation or found it off-putting. He drew on a wide range of personal knowledge to help the mathematics mean something to his students. Examples adopted almost every imaginable viewpoint: the workman who must get his quantities of material right; the factor who must manage multiple accounts dextrously; the substantial landowner who would redesign his bowling green or compute the value of his shipping interests discounted against time or loss.

Not surprisingly, it was the perspective of the merchant that returned again and again, and international trade was seldom far from view: 30 barrels of anchovies, 71 hundredweight of tobacco, 5 chests of sugar, 3 barrels of indigo. You can almost hear Hutton telling his students (and their parents): See how useful mathematics is, how rich it can make you, how much it can transform your life.

Writing a book was a much bigger step than sending in problems and solutions to magazines, and it demanded much more care. Blunders now would be costlier than wearing an embarrassing garment in a pit village. For the first edition Hutton paid for the printing himself, meaning that he alone bore the financial risk in case the book failed to sell. His patron Robert Shaftoe in fact contributed to the cost in return for the book's dedication to him. The book was produced by a local print shop, with Hutton reputedly cutting his own type with a penknife when the shop didn't have the fractions or algebraic characters the book needed.

The *Guide* was advertised in a number of newspapers, but of direct reaction there was practically none: no reviews, no comment in the press. It faced stiff competition. Even within Newcastle there were other mathematics textbooks being promoted, and other mathematical authors longer established and better known. The Banson dynasty, who dominated the city's Free Writing School, had been publishing their own arithmetic books since 1709, most recently in 1760. Another northern author had an 'easy introduction' to mathematics out in 1763.

Despite that, the *Guide* found a market. We don't know how many copies were printed, but a decent stock had sold out within a year or so, and Hutton managed to interest a London publisher in bringing out a second edition. This was good news, and ensured a much wider circulation for the book, this time at no financial risk to the author. His growing reputation was doing its work. By the time of the third edition, in 1771, the advert could say that the little book had 'been found . . . useful in schools all over the kingdom'. The *Guide*, in fact, would run and run: it was still in print in the 1860s. The name of Charles Hutton was becoming harder and harder to avoid if you were interested in mathematics and its teaching.

Contacts in London made a huge difference. After the *Guide* Hutton devised a new, more ambitious publication project: a book on mensuration. This could have been a subject for another slim textbook on the model of *The School-master's Guide*. But Hutton had something much grander in mind. Not a little book of practical rules but a veritable encyclopedia covering every aspect of geometry and its practical use. Hutton took to riding over to the village of Prudhoe at weekends to consult with the schoolmaster there, a Mr

Young, who coached him in advanced geometry and mensuration and, it was said, worked over drafts of his new book with him.

Announced in the Newcastle papers in December 1767, Hutton's *Treatise on Mensuration* appeared in twenty-eight instalments between March 1768 and November 1770. His publisher diligently promoted it in a range of national and local newspapers. Hutton undertook his own publicity campaign, writing personally to a long list of philomaths culled from *The Ladies' Diary* and elsewhere. He obtained permission to dedicate the book to the Duke of Northumberland.

The results were spectacular. When the *Mensuration* appeared as a single collected volume at the end of 1770, the list of subscribers contained more than six hundred names. Probably amounting to more than half the active mathematicians and lovers of mathematics in the United Kingdom, from Penzance to Dundee, they included two dukes, one earl, and astronomers from Oxford University and the Royal Observatory. Both the English universities and most of the Scottish ones were represented, as were surveyors and instrument makers, schoolmasters and country curates, surgeons, excise officers and Fellows of the Royal Society.

This was self-publicity on a scale rare in Hutton's century, or indeed in any. A good deal of money was involved – 600 subscriptions at fifteen shillings a book were not to be sneezed at – but the visibility had a value that could hardly be measured. It was rapidly becoming impossible to do mathematics, to like mathematics, to be aware of mathematics in Great Britain without knowing Charles Hutton's name. Hutton was well advanced on the road from provincial schoolmaster with a taste for mathematical puzzles to national celebrity. He was aware of the change himself, of course. Throughout the 1760s he called himself 'schoolmaster' or 'writing master', but by the 1770s his title pages proclaimed him 'author'; in 1772 he would switch to 'mathematician'. And while Tonthu had been 'of Newcastle', Charles Hutton could call the town coldly 'that part of the country in which I reside', implying choice, impermanence, a lack of decisive ties to his provincial life.

What the subscribers to the *Mensuration* got for their money was a fat book which, despite a title that associated it with practical matters, was in fact a comprehensive treatment of theoretical as well as practical geometry – a substitute for Euclid's *Elements*, indeed, as far as geometry was concerned. It began with definitions ('A Line is a length conceived without breadth'), and it ran all the way up to the volumes of polyhedra and the areas contained under algebraically defined curves. Most of the book consisted of increasingly complex geometrical problems with rules for solving them: to find the area of a semicircle; to find the volume of a segment of a sphere; to find the surface area of a hyperboloid. Compared with the *Guide*, the emphasis was much less on carefully graded examples and much more on comprehensiveness, on a solid, gap-less treatment of a large body of material. It explained how to find the areas and the volumes of certain shapes and solids; how to construct certain curves and surfaces such as those arising from slicing a cone, or from rotating the conic sections that resulted ('A conoid is a solid conceived to be generated from the revolution of the parabola or hyperbola about the transverse ax[is]').

There was elegance and beauty here; there was also a tremendous display of learning. Hutton regularly succumbed to the temptation to exhibit his own cleverness at the expense of relevance or logical structure. Some of the more advanced, more difficult or more important solutions were backed up by proofs given in footnotes – and sometimes the footnotes were long, elaborate, even showy. Attached to a problem about right-angled triangles he indulged in not merely a demonstration but a 'General Scholium' (the term irresistibly recalls Isaac Newton's use of it in his magnum opus, the *Principia Mathematica*) with 'some new theorems concerning the relations of the sides and angles of triangles'. These derived from infinite series – never-ending algebraic formulae – for the sine of an angle. A problem aimed at finding the circumference of a circle from its diameter and vice versa occasioned a note which set out the history of infinite series for the circular ratio from the seventeenth century onwards. Finding the length of part of an ellipse became the occasion for another, typical burst of complexity, with the full apparatus of geometry, algebra and calculus deployed to provide five different ways of solving the problem. Passages like this give an impression of extraordinary authorial dexterity: partly because they seem slightly out of place, obtruding upon the reader's notice.

So this book, like the *Guide*, carefully fashioned Hutton himself. As seen in the pages of the *Mensuration*, he was a technically dextrous, extremely well-read mathematician, who could quote thinkers of the calibre of Newton as easily as obscure practical manuals of gauging and surveying. He was entitled to expect a lot of his readers: introducing calculus without explanation, stating rules without proofs because they were 'too evident' to need it, demanding faultless skill in imagining three-dimensional shapes and their manipulations.

But this was also someone who knew all about mathematical practice. The final section of the book turned to the practical applications of geometry, and worked through lengthy rules and examples for surveying, gauging liquid volumes, and measuring roofs, windows and chimneys. Hutton went to some trouble to emphasise his experience as a surveyor; in fact his first, last and only published survey – a map of Newcastle – appeared in the same year, 1770. He suggested improvements to practice, dismissed some instruments and praised others, and casually remarked on the best way of operating in certain situations. He urged the use of decimal, not duodecimal arithmetic by measurers, and he lambasted sellers of timber for their sharp practices, at rather petulant length (Hutton must have had a bad experience on this score; he wrote a letter to one of the Newcastle papers about it too).

This practical section culminated with the imagined exercise of planning and building a house. With columns at the front door and over six hundred feet of plaster mouldings, maybe the imagined house, a smart, ambitious Georgian edifice, bore some resemblance to the real one Hutton had planned and built for himself on Westgate Street.

Not everyone was convinced by Hutton's posturing. There were some poor reviews in the London papers, by anonymous authors whose objections centred on his pretensions to practical knowledge. By and large, though, the reaction seems to have been positive. And the *Mensuration* stood the test of time; it would run to four editions, remaining in print until 1812. As late as 1830 a supporter would call it the work of a 'masterly hand', still 'by far the best treatise on mensuration' published in any country. One reader in Shropshire broke forth in verse to express his admiration:

> O Science! trade and commerce are thy end,
> By thee we import, and by thee we vend;
> By thee we build our houses, till our lands,
> And weigh and measure with unerring hands.
> What art or rules could never yet display,
> Nor all the rules of Science till this day
> Were able to disclose [by] genius' force,
> Thy true-born son hath traced to the Source.

Newcastle bridge in ruins.

Hutton's daily round of work must have been frenetic during these years in Newcastle around 1770. To teaching and administering a school, together with a continuing programme of reading and self-improvement, were now added the demands of writing and seeing through the press a steady flow of publications. Mathematics is hard to proof-read, and a geometry book like the *Mensuration* also required hundreds of diagrams to be commissioned and checked. A young Newcastle engraver named Thomas Bewick, subsequently a celebrity himself for – particularly – his engravings of birds, did some of his first work on the *Mensuration*, and he remembered that Hutton frequently came into his work room to inspect what he was doing. Despite Hutton's increasingly assertive public persona, he found him 'grave or shy' in this private setting.

Hutton remained capable of a misstep. Late in 1771 the bridge at Newcastle was partly destroyed by a flood, massively disrupting travel in the city and the area. While an inquiry was taking place and before a replacement had been decided on, Hutton rushed into print – it's hard to see how he got the text through the press as fast as he did – with a 102-page book on the theory of bridge building. It was an able, even a learned work, full of geometrical diagrams and cutting-edge mathematical analysis using some very clever calculus; the voice was recognisably the same self-confident

mathematician of the *Mensuration*. The chief problem tackled was to find what shape the upper curve of a bridge should be for a given lower curve – or vice versa – in order to ensure its theoretical equilibrium. Hutton found some neat results and expressed a decided preference for a curve of his own invention, related to the catenary (the shape of a hanging rope or chain).

All this was very well, but as the reviewers were quick to point out, few people could have been less qualified than Charles Hutton to lay down strictures about how a bridge should be built. He had no practical knowledge of structural engineering, and his mathematical analyses both contained elementary technical errors and displayed total ignorance even of the actual shapes of bridges. His formulae produced absurd results, demanding a bridge of infinite thickness in some cases. 'We are at a loss to account for the author's design in presenting this work to the public,' concluded one. If that stung, and if Hutton regretted his rush into print, his reputation by this point could stand the damage.

Another large project was already in hand, too: a cheeky scheme to reprint *The Ladies' Diary* itself. What better way to appeal to the philomaths of Britain? Hutton proposed a five-volume collection featuring every puzzle, problem, question and answer that had appeared in the *Diary* since its foundation in 1704.

It was a finely judged plan. Older copies of the original *Diary* were hard to come by – some issues extremely so – and as long as the price was right most of the hundreds of British philomaths could be relied on to purchase this collected edition of their favourite reading. Yet it would involve very little work: simply transcribing and very lightly rearranging the text from seventy slim pamphlets. Hutton even engaged an assistant to do the actual editing. (George Coughron's was an interesting story in itself, but a tragically short one. Son of a farmer, good at mathematics, he won a national mathematics competition sponsored by the *British Oracle* and moved to Newcastle, where he fell in with Hutton, perhaps a little in awe of the older man – he was twenty to Hutton's thirty-five. Their collaboration was not to continue past the *Ladies' Diary* project, and Coughron fell victim to smallpox in 1774.)

Hutton added more extra matter to his *Diarian Miscellany* than might have been expected. It was issued in quarterly 'parts', and to each he appended a selection of new mathematical essays and correspondence. In true philomath style he also included a set of questions each time: readers' answers to be printed in the following issue.

Hutton's associates Coughron and Fryer themselves made contributions, as did a range of big names in the philomath world and a number of other people who were probably Hutton himself: 'Nauticus', 'Geometricus', 'Astronomicus' and 'Analyticus'. He felt confident enough to admit to some errors in his *Mensuration*. But the problems column never really took off, perhaps because the market was already so crowded.

Nevertheless, Hutton found plenty of buyers for the *Miscellany*: it came out in slim 'parts' between July 1771 and July 1775. His publisher once again pushed the project hard, taking out advertising space in the national papers for every one of the thirteen separate parts.

So good was the idea that someone tried to copy it, and Hutton found himself with a rival. Samuel Clarke launched a *Diarian Repository* in 1774, printing just the mathematical items from *The Ladies' Diary* and leaving out the other puzzles, and there was an undignified flurry of adverts and counter-adverts in the press. Hutton thought the work defective and took to calling it the 'Repository of Errors', but he was badly rattled by the incident, and matters were not helped when a couple of the numbers of his *Miscellany* were delayed at sea by bad weather; printing in Newcastle for distribution from London had its perils. But Hutton won the day. Clarke's *Repository* abruptly ceased to appear, some months before reaching completion.

When Hutton's *Miscellany* was printed complete in 1775, the collection made a handsome set of five neat volumes for which the publisher asked one pound nine shillings. The new matter took up an extra volume by itself, and was separately titled – a little pompously – *Miscellanea Mathematica*. Well indexed, the *Miscellany* became for some decades the standard reference for *The Ladies' Diary* and its problems and contributors, and it incidentally contributed much to the *Diary*'s accessibility to modern historians.

∞

Up to this time – the early 1770s – it's not clear that Hutton's relentless self-promotion in print had any very definite object. There were a number of relatively prestigious mathematical appointments available in Great Britain, including teaching jobs at institutions like the Royal Mathematical School at Christ's Hospital, or the public lecturing posts at Gresham College in London. The chairs at Oxford and Cambridge were not realistically open to someone without a university degree, but a position in one of the ancient Scottish universities might have been possible with the right use of Hutton's personal network. If he explored any of these possibilities, no trace of it has survived.

What happened in fact was a little more surprising.

∞

Hutton was in London on *Mensuration* business when he heard from a friend – Edward Williams of the Royal Artillery – that an unusual job was open. John Lodge Cowley, professor of mathematics at the Royal Military Academy at Woolwich, had retired due to declining health. The job attracted two hundred pounds a year and a house at the Academy. Plenty of people wanted it, and quite probably some of them knew some mathematics. The official in charge – George, Viscount Townshend, Master-General of the Ordnance – reckoned it would be a good idea to fill the post with a man of ability rather than a well-connected nonentity. He had taken the rare step of announcing a public competition, giving it to be understood that 'merit alone' would decide the result. Advertisements had appeared in the newspapers.

Williams urged Hutton to present himself as a candidate. Back in Newcastle his supporter Robert Shaftoe did the same, and Hutton let himself be persuaded. In May 1773 he travelled back to London to take part in the supposedly impartial competition.

Merit alone? 'Merit is useless,' wrote Horace Walpole, 'it is interest alone that can push a man forward. By dint of interest one of my coach-horses might become poet-laureate, and the other, physician to the household.' There were increasingly meaningful public examinations at the British universities during the eighteenth century, and it was not absolutely unknown for public appointments to be made on a competitive basis. Expectations about public life and the right use of patronage were changing, after all, but they were changing slowly.

So the candidates for Townshend's 'impartial' examination turned up armed with the usual array of recommendations from noblemen and politicians: testimonials, promises of favour or reminders of favours owed. Hutton would later claim that he himself competed 'without any interest'. But in reality he too took his precautions. He obtained a recommendation from his acquaintance the Duke of Northumberland, to whom he had dedicated his *Mensuration*. Northumberland was a prominent Tory, a colliery owner and a Fellow of the Royal Society, and he controlled seven votes in Parliament. Through Robert Shaftoe, Hutton also obtained a recommendation from the Earl of Sandwich, a Whig (the party was in power at the time) who had been First Lord of the Admiralty; to Master-General Townshend his was probably a louder voice than

Northumberland's. Having covered both political bases, Hutton finally equipped himself with a letter from the prominent northern mathematician William Emerson, testifying to his intellectual abilities. Thus protected, he faced the examiners.

Theirs were names he knew, but none was even an acquaintance. Three, however, were men you had almost certainly heard of if you had any contact with the British scientific and mathematical scene. Nevil Maskelyne was the Astronomer Royal; Samuel Horsley a member of the Council of the Royal Society; John Landen a highly regarded mathematical author and Fellow of the Royal Society. The fourth examiner, representing the military side, was Henry Watson, a military engineer with experience from Havana to Bengal as well as a contributor to *The Ladies' Diary* and a friend of and literary executor to its former editor.

They started by examining the candidates viva voce. It was a stiff competition. Other candidates included Benjamin Donne, Master of Mechanics to the King, and Hugh Brown, translator of major works on military mathematics. (There were at least six candidates in all, but Hutton's telling of the tale seems to have grown over the years. The obituaries have him facing down a field of nearly a dozen.)

The questions ranged across the entire field of mathematics, its history, what the best books were on certain subjects, and how best to teach it. When the first day was over the examiners handed the candidates a set of deliberately abstruse written problems in mathematics and natural philosophy, telling them to come back at the end of the week with whatever answers they could manage.

A persistent rumour – it was still circulating in 1825, after Hutton was dead – said that Hutton, away from his books and his friends, struggled with the written questions, and visited the Duke of Northumberland in London in some despondency. The Duke, so the tale goes, got the examiners to provide a new, fairer set of written questions. Even if that is true, it must have been a tough week, wrestling with the problems all day for several days before finally writing up whatever answers one had been able to obtain and turning them in at the end of the week. And Hutton must have spent a tense weekend in his London lodgings waiting for the result.

It came on Monday. The examiners reckoned that most of the candidates were sufficiently well qualified to do the job and had given complete satisfaction as far as the written and verbal questions were concerned. But they felt compelled to single out Hutton for particular recommendation, on account of the exceptional strength of his performance. The Board of Ordnance ratified their decision on 25 May. And so patronage and merit worked together (as they sometimes did) and Charles Hutton became Professor Hutton, of the Royal Military Academy at Woolwich.

For Hutton it would be the biggest single change since he left the coal pits. And it all happened with furious speed. In less than a fortnight the school in Westgate Street was to let; in six weeks it was taken by John Fryer, Hutton's former assistant. He advertised that he intended to keep up the teaching 'in the same Manner as practised by Mr Hutton', and had hopes of retaining Hutton's students as his own.

And Hutton himself, after a brief visit to put his affairs in order, was gone from Newcastle. He would never visit the region again.

The coach journey – three days and two nights – had nearly shaken him to pieces on previous trips, so this time at least he made the journey to London in the different style offered by sea travel. It might take a couple of weeks, and he perhaps reflected that this was the same route that took so much of Newcastle's other produce to the capital. Almost a million tons of coal sailed from Tyne to Thames annually, the ships crowding the two rivers. In June 1773, so did Charles Hutton.

4

Professor

Woolwich. South of the Thames, east of the City of London. The 'Warren'. Rabbits were bred there in the Middle Ages, and by 1773 it's an apt name once again. The teeming site is now Britain's biggest munitions manufactory, its largest ordnance store.

Five hundred people working in a hundred acres. The closest thing in the Georgian world to a modern factory. Warehouses, workshops, laboratories, furnaces. A canal, a barracks, a parade ground. The Royal Artillery is quartered there; so is its cadet company. So is its band.

It's loud; it's dirty, dusty, smoky. The smells of gunpowder, its smoke and its ingredients hang heavy in the air: sulphur, saltpetre and charcoal. A harsh, confusing landscape, security-conscious and alienating to outsiders. Not so different from the collieries, perhaps.

There were two sides to the change in Charles Hutton's status. He was no longer a member of that despised class the provincial schoolmaster, with a merely de facto high standing among British philomaths. His status was no longer based solely on his own furious efforts at self-promotion. He was no longer vulnerable to catastrophe the moment the fashion in Newcastle schools changed or a rival published a cheaper textbook. He was now a man who had received public recognition for his talents and his hard work, in the form of a teaching appointment in an institution of national significance.

On the other hand, Hutton had left a situation as absolute lord and master of a thriving business and had become an employee. He had changed from headmaster to ordinary teacher, and, what was more, he was becoming a civilian employee in a military establishment. He would have to work with a new set of colleagues and teach a new kind of pupil. He would have been less than human had he not experienced moments of doubt about what he was getting into during the summer of 1773.

First impressions were certainly discouraging, and in some ways Hutton retained mixed feelings about his new situation for years, as his letters show. The Royal Military Academy occupied a few rooms in the bustling military installation at Woolwich, downstream from the city of London and on the opposite side of the river. Docklands territory today, it was then barracks, ordnance factory, munitions depot and more. The dockyard went back to Henry VIII's time, ordnance testing over a century. There was a laboratory for making gunpowder and a foundry for making ordnance and shot. By the mid-eighteenth century the ever-increasing collection of buildings and activities was straining the limits of possibility. Laboratory, arsenal stores and ordnance testing were much too close together, and from time to time there were fires or unplanned explosions. The latrines stank. There was too little water to go around. The four battalions of the Royal Artillery would move out of the site in 1777, but the Academy – despite sporadic complaints and a serious attempt to find a new site in the 1780s – would remain at the Warren until the next century. Contemporary guide-books tried to talk the place up, with varying success:

The Royal Military Academy at Woolwich.

In the warren, or park, where trial is made of great guns and mortars, there are some thousand pieces of ordnance for ships and batteries; with a prodigious number of shot, shells, and grenadoes, heaped in large piles of various forms, and which have a very striking and pleasing appearance.

Even today, with the mounds of weaponry gone and the site sleepy, and with a good map in your hand, it takes an effort to find the building in which the Royal Military Academy held its schoolroom. The two imposing floors by Vanbrugh had windows east and west to let the light all the way through, with one big high-ceilinged room on the top floors. There were also a barracks for the cadets and dedicated houses for the two principal masters, of whom Hutton was now number two. His house needed ten pounds' worth of repairs, hastily carried out during the summer he arrived.

The human landscape was not much more encouraging. Hutton's family didn't at first accompany him to Woolwich, and his new colleagues were a somewhat sorry crew. The Academy had existed for over fifty years, with a formal king's warrant since 1741, but it was proving hard to get the setup right. Reform after reform had taken place. Initially it was meant as a sort of regimental university, providing lectures to anyone in the Royal Artillery who chose to attend: officers, NCOs, cadets, all those associated with the regiment. A series of mid-century reforms had given it a more manageable role as a sort of superior school for the Royal Artillery cadet corps. There was now an inspector-of-studies, with military rank, to look after the daily running of the Academy, to oversee the curriculum and make sure it was being delivered. Above him stood the Lieutenant-Governor, who reported to the Board of Ordnance and its Master-General. (The Ordnance, by the way, was not strictly speaking part of the Army: it was structured and governed separately, which goes some way to account for the uniqueness of its Academy.)

Some of the teaching staff had never accepted the changes to the style of the institution, or the inspector's attempts to regulate what was taught and how it was taught; possibly they preferred the idea of themselves as university lecturers to that of school-teachers. The chief master, Allen Pollock, had all but declared war on the management, doing everything he could to make the inspector's task difficult. In principle he taught the prestigious subject of fortification and artillery. In practice he increasingly tended to teach absolutely nothing. He lived miles away from Woolwich without ever receiving permission to do so, and it was his habit to appear literally hours late for his lessons, then waste more time shuffling his books and his papers. For every adjustment to his teaching duties he demanded a direct written order.

There were also masters for French, dancing, fencing and 'classics, writing and arithmetic'. The first two struggled to command the respect of the cadets – to put it delicately – while the last, a Reverend William Green, followed Pollock's lead in calculated awkwardness.

Gentlemen Cadets.

Hutton would have learned at least some of this over the summer, while his new house was being repaired and he was finding his feet after the upheaval of his move from the North. When the boys arrived after the summer recess and the actual teaching began, it may have been something of a relief. The cadets, numbering about fifty, were a well-dressed body of young men with a rather splendid uniform: blue coats trimmed with scarlet and gold; lots of lace, hair neatly queued and powdered.

They may have looked genteel, but they ranged in age from twelve to about nineteen, and there were predictable problems with their behaviour. Some ingratiated themselves with the masters, sending home for hams, hares, pheasants and other little gifts that might smooth their passage through classes and exams. Others indulged a tendency to riot. Orders and letters from the governor remarked again and again on matters of discipline, expressing futile outrage and lamely prohibiting misdemeanours after they had taken place. Cadets must not ill-use the masters, must not treat them with disrespect or insult them. They must not disrespect their superior officers, nor use indecent or immoral expressions. They must not wilfully destroy fixtures, bedding, furniture, utensils or windows, nor must they force locks or open or spoil the desks of the teachers or other staff. They must not 'nasty the walls or chimneys'. They must not leap or run over the desks, nor climb out over the walls at night.

As the crimes escalated, so did the punishments. ('The first Cadet that is found swimming in the Thames shall be taken out naked and put in the Guard-room.') We read in mid-century accounts of solitary confinements, bread-and-water diets, imprisonments in a dark room, degradations and a few exemplary expulsions. In the long run, things improved, but outbreaks of quite serious violence remained more frequent than anyone could have wished. The constant presence of a duty officer in the classrooms had little effect. There was continued noise, hallooing, shouting; there was window-smashing, and 'pelting' of the masters. A lieutenant lost the use of his middle finger. A cadet spent three years on the sick list after a prank whose details he permanently refused to specify but which he steadfastly described as a piece of 'fun'.

Charles Hutton was suddenly a very long way from the genteel sons and daughters of middle-class Newcastle: the grammar school boys and girls taking advanced mathematics lessons; the private pupils paying a few guineas extra for dancing classes and a few more for the course of lectures on astronomy. Yet his orders required him to teach the cadets, and to teach them a demanding curriculum that ranged across pure and applied mathematics. Teach it he would. The theory classes were on Mondays, Wednesdays and Fridays. Professor Pollock – if he bothered to turn up – taught the boys his notions of fortification and artillery for three hours in the morning; in the afternoon Hutton took them for mathematics from three till six.

Facing about twenty-five of the older or cleverer boys, grouped at least notionally into four graded 'classes', he covered Euclid's *Elements* of geometry, killing two birds by using algebraic explanations wherever possible. Trigonometry as applied to fortification and measuring, such as finding inaccessible heights and distances, and using logarithms for the calculations if needed. Conic sections (the shapes you get when you slice a cone using a plane). Mechanics with reference to moving heavy bodies (such as pieces of artillery): levers, pulleys, wheels, wedges and screws. The laws of motion with reference, naturally, to projectiles. Calculus – 'fluxions', in the Newtonian rather than the Continental style – and whatever uses could be found for it. Much of this would have been familiar stuff from his Newcastle classroom, but the style of teaching still made it an exhausting round.

As in any other Georgian classroom, the boys were mostly at their desks, called forward one at a time or in small groups. Each boy did some exercises, showed them to the teacher, copied them and the relevant section of the textbook out fair: and repeated that round for the whole three hours. For Hutton it might mean fifteen minutes with one cadet on calculus; fifteen minutes with another on Euclid. Then algebra, then logarithms, then something else. He also gave periodic lectures – virtually the only university-style lecturing still taking place at the Academy – on geography and the use of those favourite pieces of eighteenth-century teaching apparatus the terrestrial and celestial globes: two hours every other Wednesday.

When they weren't in the theory classroom, meanwhile, the boys were instructed on the other days of the week in the practical arts of gunnery: loading and firing the artillery pieces, and pointing them in the right direction; exercises in hitting marks. Various other practical matters: transporting burdens by making pontoons, floats or bridges; mortars and their use; trenches and mines: how and where; the composition of gunpowder; the names of the parts of a gun. Casting and weights of artillery pieces and shot. Handling of stores.

As if that wasn't enough, from 1778 a separate building was set up as a 'military repository'. Here the boys were taught the finer points of handling the equipment: mounting and dismounting ordnance, crossing obstacles, and 'overcoming by resourcefulness difficulties which did not arise in the ordinary routine'.

As far as dangerous fun for boys went, this was infinitely more attractive than the classroom work, and the cadets were after all in training for a severely practical life as officers in the artillery or in some case the Engineers' corps. On graduation at about eighteen most of the boys were at once commissioned as second lieutenants in the Royal Artillery regiment.

This, then, was the real stuff of an artillery officer. Cannon up to ten feet long and weighing up to three tons; iron balls moving at the speed of sound. Howitzers; mortars for lobbing shells high overhead. Roundshot, grapeshot, case-shot; explosive shells. The relentless drill of firing: four men at work, and every action had to be right every time or disaster would follow. Loader puts the charge down the muzzle, in its paper cartridge. Ventsman blocks the vent with his thumb, and spongeman rams the charge home. Ventsman pricks the cartridge, primes it with quickmatch in a tube pushed down the vent. Loader puts the projectile in the muzzle, spongeman rams it up against the charge. Run the gun up, point it at its target: traverse, align, adjust the elevation. Firer touches the vent with his portfire. The blast of several pounds of gunpowder, and an iron ball travels a thousand yards. Check the recoil. Spongeman sponges the barrel clear. Repeat. Furious speeds were possible; with a light gun seven rounds a minute could be achieved in an emergency: nine seconds from one shot to the next.

In alternate years the summer saw a full-scale attack upon a polygonal front built at Woolwich for the purpose, done 'with all the form and regularity that is used in a real siege'. Trenches, batteries, mines made by the besieged to blow up the batteries, mines made by the besiegers to make breaches in the walls. Fortunately there was a well-equipped hospital on the site.

With all this going on, it may be no wonder there was a problem getting the boys to attend to their theoretical studies. Some of them at least itched to get away from the mathematics and the theory, and there were still in Hutton's day at least some officers of the regiment who thought the whole Academy a waste of space. General Belford wished it disbanded, and in the late 1770s wrote a stinging letter to the Master-General of the Ordnance suggesting that instead 'a number of fine young fellows [be] appointed as Cadets to every Battalion', where they might do practical work and see real action.

He may have had a point. The long series of reforms and adjustments had made the Royal Military Academy an institution whose details were not easy to defend. Admittedly some kinds of knowledge were best acquired in a school-like setting, but why should an artillery officer need to know algebra, geometry, trigonometry?

But the training was not so much about what you needed to know as about producing a particular kind of man with a particular kind of mind. The Academy, unique and exceptional in some ways, was nevertheless part of a much more general eighteenth-century trend in English education. Mathematics and the sciences enjoyed increasing prestige at the English universities, with Cambridge in particular stuffing its curriculum full of Euclid and Newton to the oft-remarked exclusion of almost anything else. Newtonian mathematics, together with the philosophy of John Locke and the rational theology of such as Samuel Clarke, 'went hand in hand through our public schools and lectures', said one contemporary.

Cambridge University's official calendar, too, boasted specifically of the 'many excellent lectures in mathematics' delivered in the colleges. John Locke himself had written that mathematics was 'a way to settle in the Mind an habit of Reasoning closely and in train . . . In all sorts of Reasoning, every single Argument should be managed as a Mathematical Demonstration.' It enlarged the mind, instilled intellectual humility, and taught the habit of clarity in one's ideas.

It was a similar picture at the so-called Dissenting Academies. These were advanced schools or mini-universities for those whose non-Anglican commitments excluded them from Oxford and Cambridge. By mid-century there were dozens of them, and they enjoyed a fine reputation, often teaching curricula based on mathematics and natural philosophy as well as logic and theology, sometimes with the specific intention of disciplining the mind, teaching logical thought and a reliance on pure reason.

So the Royal Military Academy was doing nothing very odd by filling its curriculum with more mathematics than any practical consideration demanded. It was doing what its managers and staff conceived as its clear duty: training and moulding young men in desirable mental as well as physical habits. Just as the duties of a gentleman officer required a knowledge of dance and a smattering of French, they required an acquaintance with the results and, more importantly, the methods of Euclid and Newton. It was a situation that suited Charles Hutton; in any debate about whether it was worth teaching boys geometry, algebra and the rest, there was no doubt which side he would take. Over the years he would become ever more committed to the cause of promoting mathematics, its usefulness and its beneficial effects on the mind. Indeed, the status and the national importance of mathematics would exercise him for much of the rest of his life in one forum or another.

As a teacher, meanwhile, he seems to have succeeded in catching

the attention of the boys, and in winning them over to take seriously what he had to impart. His manner was cheerful, friends
reported, but deliberate in expression; he himself reckoned gravity
a part of his character. His voice was clear and firm, with a slight
northern accent that he would keep to the end of his days. His
mind (in good Lockean style), they said, was 'accustomed to commu-
nicate its feelings with a sort of mathematical precision'. Hutton
was remembered by his former students with respect, affection,
even veneration. One cherished a commemorative medal of Hutton's
august profile to the end of his life. It was never Professor Hutton
who got pelted in the street.

∞

If the boys were won round to Hutton fairly rapidly, the staff were
a tougher task. Pollock was going from bad to worse. The teaching
staff were civilians; so, unlike an officer of the regiment, he couldn't
be court-martialled for disobedience. Reprimand after reprimand
had no effect on his attendance or his willingness to deliver the
course as the inspector conceived it: just more elaborate, sometimes
legalistic excuses. He disputed the rules about where he was
supposed to live, and the system by which certain parts of his pay
were worked out. He disputed what he was supposed to be teaching
and where the boundary lay between Hutton's subject area and
his own.

There was indeed a long-running question as to whether the
curriculum should be determined by the Board of Ordnance
(which paid for the Academy) or by the teachers who had to
deliver it. And there was a more detailed question about how
responsibility for the practical parts of mathematics – surveying
and so on – should be divided between the Professor of
Fortification and Artillery (Pollock) on the one hand and the
Professor of Mathematics (Hutton) on the other. In both respects,

sane compromises took some time to arrive at, and Pollock made himself increasingly ridiculous. Early in 1774 the inspector wished the more advanced boys to be shown how to take an angle of elevation with a theodolite or quadrant, and asked Pollock either to do it or to lend the instruments to Hutton. He refused 'in a very haughty and imperious manner' and added 'that the Academy is not a fit place to mention those things'. He took to refusing to allow the inspector into his classroom, standing on a technicality in his written instructions.

Hutton as Professor of Mathematics also shared a curricular boundary with the Master of Classics, Writing and Arithmetic: William Green, another difficult character. For three days a week, while Hutton and Pollock were teaching the older boys, Green taught elementary mathematics and writing – and possibly some Latin – to the younger and less able. He received a smaller salary than Hutton and Pollock and taught in theory twice as many hours as they did, so he had some reason for ill feeling. And his duties were subject to repeated redefinitions during the 1770s, the teaching of classics being in part suspended at some periods and the teaching of writing theoretically unnecessary since illiterates were not supposed to be admitted as cadets. Harassed and dissatisfied, Green became irregular in his attendance, and eventually his permission to live away from Woolwich was withdrawn. His performance affected Hutton directly, since if Pollock taught the boys badly they would graduate to the upper academy, Hutton's domain, inadequately prepared.

In this somewhat chaotic situation, Hutton could have thrown in his lot with the masters: done minimal work, obstructed the inspector and governor in their duty, and hoped to get away with it for as long as he could; although by the early 1770s it was tolerably obvious that dismissal was on the horizon for the erring Pollock. Or he could take the other path.

It wasn't a hard choice, and indeed it wasn't a hard task. In a way the situation was a gift to him. Merely by turning up regularly (and he was a punctual man) and actually delivering the nine hours of instruction he was being paid for each week, Hutton necessarily outperformed Pollock and Green – incidentally making them look worse than they did already. They resented it, but they could hardly stop him. The inspector was won over quickly, and Hutton was rewarded with a growing degree of power and influence that went beyond anything in the written instructions. There began the first of a very long series of organisational changes at the Royal Military Academy that reflected Hutton's priorities and agenda.

Early in 1774 an entrance exam was instituted in consultation with the inspector and masters. Pupils should be 'well grounded in the first four rules of Arithmetic, with a competent knowledge of the Rule of Three', as well as some Latin grammar. It's hard not to see Hutton's hand in a change that would ensure the boys he had to teach weren't utterly incompetent in his own subject. In theory the curriculum stood still, but there is no doubt Hutton was adjusting it to suit his notions, and by 1775 his *Mensuration* was a set text.

And, crucially, by the late 1770s it was at Hutton's word alone that boys graduated from the lower to the upper academy. For the boys, graduation was a desirable thing in itself: a move from learning basic arithmetic and basic military drill to being taught advanced material by the two professors, introduced to the properly military subjects of artillery and fortification, and allowed to wear the full uniform and sword of the cadet company. Furthermore, promotion within the Royal Artillery and the Engineers' corps – like the Navy but unlike the Army – was entirely by seniority, which meant that the date of passing one's exams relative to the other cadets was highly significant, setting to some extent the course of an entire future career. With Hutton's judgement now vital to the progress of cadets through the Academy, he became in some ways a very important man indeed.

Meanwhile, in 1777, Professor Pollock was pensioned off with an almost insulting fifty pounds per year. He tried to make one final fuss, but the Board of Ordnance wisely refused to be drawn into any argument about the matter.

Soon afterwards William Green brought to a head his resistance to what he evidently felt was an alliance of management and Hutton. He disputed Hutton's academic judgement and demanded that a Mr Mudge, and certain other boys, be moved to the upper academy despite Hutton having failed them. He stated that they were more worthy than some who had been moved up, that they had solved several questions in algebra that members of the upper academy were unable to understand. He spoke of 'much injustice' and named several cadets.

This was a key moment for Hutton at the Academy; if Green's complaint had been upheld Hutton would have looked absurd and his position would have been scarcely tenable. Perhaps not surprisingly, a board consisting of the governor, the inspector and the new Professor of Fortification and Artillery found 'that Professor Hutton has done justice'.

After this vindication there was no doubting that Hutton was the de facto academic head of the Academy. In time the Professor of Mathematics ceased to be called 'second master' and became the first. There is no further report of questions about his academic judgement, or of challenges to his right to pass cadets from one class or academy to another. The notion that the teaching of mathematics and mathematical competence were at the heart of the Royal Military Academy had triumphed. Green, perhaps surprisingly, was brought round to accept Hutton, and the two worked together with no more outbursts until the older man retired in 1799.

∞

Pollock's departure, replaced by a congenial Frenchman named Landmann, might have marked the commencement of a period of calm for Hutton's work at Woolwich. But the mid-1770s saw rather more than just a few schoolmasterish tensions come to a head. In December 1773 the Bostonians dumped ten thousand pounds' worth of tea into their harbour. By April 1775 there was fighting in Massachusetts, and by 1778 the conflict was a global one; France became involved that year and Spain the next, making India, the West Indies and Central America into theatres of war. Hutton and his colleagues, suddenly, were training and approving young men for active, urgent deployment around the world, and their work took on a new character as a projection of British power across the globe.

Sixteen companies of the Royal Artillery participated in the American War of Independence. When Spain besieged Gibraltar in 1779 it created a particularly acute need; the siege was by its nature an artillerymen's affair and five companies of the regiment were there from start to end. The Royal Artillery would receive a special message from the King when it was all over.

In consequence, the war created a sudden need for more graduates from the Royal Military Academy. And in a pattern that would be repeated in later conflicts, the institution coped poorly. Almost at once the shortage of suitable candidates became a matter of official comment, and the Academy came under intense pressure to increase its throughput of young men at almost any cost.

Public examinations were discontinued, replaced by private examinations in front of the governor, the inspector and the two professors. Boys were examined who had never been formally admitted to the Academy or the cadet company. The intention was naturally to pass as many as possible, but by 1780 it was being remarked that cadets were being hurried through the upper academy too fast and were graduating little qualified to hold commissions. Reports from one examination coyly stated that the cadets understood 'a little Algebra, and a little Geometry'. But the expertise needed to handle ordnance effectively or even safely could not be created out of nothing just because need pressed. Permission was given for some notionally qualified second lieutenants to stay on at Woolwich for a further year in order to complete their studies.

Even in these conditions, when cadets were being pushed into and out of the upper academy with the minimum of ceremony or even propriety, the Master-General took time to note that new boys in the upper were on probation to Hutton. They were 'neither to have their full uniform nor the allowance of one shilling a-day pocket-money until admitted by the Professor of Mathematics'. He was expected 'to turn them back into the Lower School' if they displayed insufficient diligence.

To add to the disruption, Green was asked to help out by doing some teaching in the upper school, delivering quadratic equations and practical geometry to those who needed it. The lower-academy students he would normally have been teaching were presumably neglected as a result, compounding the problem of students entering the upper inadequately prepared. In 1782 an assistant mathematics master was belatedly added to the strength in a more robust attempt to fix the situation, though it remained the case that the quantity of mathematical instruction being delivered was large in relation to the number of mathematical staff. At the same time a pay rise was awarded based on the number of days the different members of staff actually taught: it had the striking effect of making Hutton and Landmann, the new Professor of Fortification, better paid than either the inspector or the governor.

Discipline wasn't improved by the disruption, and it made things no easier that the firm-handed Inspector Pattison was promoted to major general and posted to America in 1777. During the war years certain cadets were degraded for being 'in liquor' and the corporals of the cadet company collectively turned bad, threatening boys who had the temerity to outperform them academically. Cadets had to be forbidden to read 'books of entertainment and newspapers' during lessons.

Then in 1780 a rumour went around the boys that you would get a commission sooner if, rather than waiting to graduate into the Artillery, you got yourself thrown out of the Academy and had your parents obtain a commission in an Army regiment by purchase or interest. Result: a spate of boys courting expulsion. France having recently entered the war, the French master was their target of choice. Cadets threw stones at him while he was trying to teach and continued to pelt him with dirt and stones on his way home. The size of the stones is not recorded, but the incident had the character of a serious assault, not a prank. Those in charge at the Academy acted with unexpected wisdom. They duly expelled the leaders of the assault, *then pardoned them.*

Another indirect consequence of the American war was that convicts could no longer be transported to the North American colonies. From 1776 they were held instead in hulks moored in the Thames off Woolwich: three ships holding nearly two thousand men. Escapes were not uncommon, adding to the woes of the Woolwich site; on occasion gun battles ensued on shore before the convicts were recaptured.

The physical situation in Woolwich remained demoralising, not to mention unhealthy. Although the Royal Artillery itself moved to new quarters on Woolwich Common soon after the outbreak of war, the Warren site was still crowded with the ordnance and munitions installations and the cadets and their Academy. Water came from a conduit house in the superbly named Cholick Lane; by the later part of the century there was too little of it to go around. One officer received permission to move house on account of the 'Horrid Smells' from the latrines.

To the chaotic situation at Woolwich was added for a while an air of national panic: in the summer of 1779 a Franco-Spanish fleet was at sea with the intention of invading Britain. That threat came to nothing, but panic was replaced in the longer term by national demoralisation. As poor strategic planning and steady underestimation of the American forces took their toll, it became increasingly clear that coercive military action was not going to solve a problem essentially political in its nature: that the seceding colonies would not be forced into submission, and that their independence was an accomplished fact. By 1780 Britain was isolated against America, France, Spain and the Netherlands, and the British public was losing its sense of why the conflict should be prolonged. The famous defeat at Yorktown in October 1781 was decisive for political as much as for strategic reasons.

When the Peace of Paris came in September 1783, Charles Hutton had been at the Royal Military Academy for ten years. It had been a draining period for the Academy and everyone connected with it, and although Hutton had established a strong position in the institution, it had cost him much in effort and exhaustion. Prone to lung disorders, he also developed chronic headaches, and he took to walking on the Academy's roof where the air was fresher. From there you could see the shipping on the river, and, to the south, the open country of Shooter's Hill and Woolwich Common. You could also see the City of London, and dream of all that it afforded.

5

Odd-Job Man

Woolwich: the Royal Military Academy. Charles Hutton's study. 13 January 1779. He's in a gloomy mood.

'I am here almost as recluse as a hermit,' he writes to a northern acquaintance, 'being almost single in my studies & manner of thinking in this place, my nearest neighbour in these respects being the Astron. Royal at Greenwich with whom I have the honour to be on very good terms.'

∞

As recluse as a hermit? What about his family?

Hutton's wife and children did not accompany him to Woolwich. His relationship with Isabella had broken down some time after the birth of their fourth child in 1769, and she never left the North-East.

Hutton and his family were always coy about this, and most of his obituarists and early biographers did all they could to avoid telling the story; so it's hard to say just what really happened. One Newcastle historian was indiscreet, though, and related in the 1820s that Hutton was initially accompanied to Woolwich by a different woman: an officer's widow named Maxwell. Apparently he soon found himself obliged to dismiss this lady on account of her extravagant habits.

That sounds slightly fanciful, and it was written many years after the fact by a writer whose sources of information are far from clear. It is certain, though, that within a few years Hutton was living with another woman. At least some of his friends knew Margaret Ord as 'Mrs Hutton' – although his first wife was undoubtedly still living – and in 1778 she bore him a daughter, Charlotte Matilda.

Who was Margaret Ord? Born around 1752, she was in her twenties during Hutton's first decade in Woolwich: eighteen years younger than Isabella. It would be interesting to know something about her background: how she compared with the first Mrs Hutton in terms of social rank, for instance. Unfortunately for the curious historian, Ord was not an uncommon name. There were Ords in the Royal Artillery and a prominent family of the same name in Newcastle; the latter, indeed, were part owners of the Long Benton colliery where Hutton, once, had worked. There were three Fellows of the Royal Society named Ord around this time. Margaret could have been related to any of them; the fact is that we know nothing of her origins, nothing of where and how she and Charles Hutton met.

Isabella remained in Newcastle and took to styling herself a widow, but by March 1776 Hutton's son Harry was in Woolwich as a cadet at the Royal Military Academy. He graduated about a year later and went into the Royal Artillery. Hutton's other children initially stayed with their mother, but by the early 1780s they too had moved to Woolwich. Just what had passed between them, their father and their mother we will never know.

Not quite a hermit, then: by the end of his first decade at Woolwich Hutton headed a household consisting of himself, Margaret, and his four daughters ranging from twenty-one-year-old Isabella down to six-year-old Charlotte.

But still, Woolwich was not London, and it could indeed feel isolated. You can walk to Woolwich from the City of London, but it takes half a day. You can shorten the time by riding, or take a boat; to row from the Tower of London down to Woolwich took a couple of hours. In time, Hutton took to renting a set of rooms in the city, at one of the Inns of Court, and spent a couple of days there every fortnight, judging that that made the best use of his time.

Much of Hutton's time was in fact his own, since his teaching filled only three afternoons each week. He took every opportunity he could find to fill that time by doing extra work and making new professional connections. He was, and would always remain, a superb networker.

Nevil Maskelyne.

A couple of miles up the river, at Greenwich, lay the Royal Observatory. This was the domain of the Astronomer Royal, Nevil Maskelyne, who was a member of the committee that examined the candidates for the Woolwich job and selected Hutton. And as Hutton put it in his letter of 1779, Maskelyne was his closest scientific neighbour. As early as September 1773 Hutton was corresponding with Maskelyne's assistant Reuben Burrow, lending books back and forth, and soon Hutton was on good terms with the Astronomer Royal himself. They would remain close until Maskelyne's death; an obituarist reckoned Hutton among Maskelyne's 'most intimate friends'.

Honest and popular, Maskelyne was a key member of the London scientific world. By the mid-1780s Hutton was sending him drafts of his papers to look at, and on occasion detailed comments from Maskelyne found their way into the published versions of Hutton's books. Hutton acknowledged Maskelyne's 'generous advice and assistance' with a dedication to him in 1785.

One of the projects in which he involved Hutton arose from his role on the Board of Longitude. The board existed to assess – and potentially to reward – schemes for finding the longitude at sea, a problem whose unsolved state was leading to losses of life for Britain and for every nation engaged in more than coastal seafaring. Maskelyne had a scheme of his own for finding the longitude: to use the moon's predictable motion across the background of stars – or relative to the sun – as a sort of clock. Starting in 1767, with the blessing of the Board of Longitude, he oversaw the printing of tables of the moon's position in the sky up to several years in advance, under the title of *The Nautical Almanac and Astronomical Ephemeris*. If you observed the moon's position and compared it with the table of predictions, you could deduce exactly what time it was. Knowing the exact time, an accurate look at the apparent position of the sun or the stars would tell you where you were.

The annual books of tables cost two shillings and sixpence; you also needed an instrument for observing the moon – a 'Hadley's quadrant' costing eight pounds or so – and a two-shilling book of extra tables. The calculations could be reduced to a feasible, if laborious, recipe that took about half an hour. The *Nautical Almanac* was distributed at ports around Britain, Europe and America, and during the final third of the eighteenth century 'lunars', so called, became an accepted method of finding your position at sea, and much the cheapest. An alternative way to determine the exact time and hence your position was to carry a really good clock; but a clock accurate and reliable enough – like the chronometers built and promoted by John Harrison – cost dozens of guineas, and unlike books of lunar tables they broke if you dropped them. Maskelyne took some criticism for his suspicion of the chronometer method, but frankly he was right; for most sailors it was still an inaccessible and impractical answer to the 'longitude problem'.

An issue of the *Nautical Almanac* contained the moon's position for every three hours, night and day, of the whole year. Making the tables in the first place was laborious, and far beyond the power of one person, even if Maskelyne had had nothing else to do (he did) and had been paid to work on the *Nautical Almanac* (he was not). Instead he outsourced the work on the cottage-industry model, to a network of human 'computers' around the country. Maskelyne's computers were teachers, surveyors, minor mathematical authors: much the same kind of people who contributed to philomath journals like *The Ladies' Diary*. Indeed, they were sometimes recruited directly from the ranks of the *Diary*'s problem-solvers.

They worked not with the equations and geometry that described lunar theory, but rather from a set of computational instructions prepared by Maskelyne. Calculating a single lunar position typically involved looking up about a dozen figures in printed tables and carrying out a similar or larger number of seven- or eight-figure arithmetical operations, all done in base 60. It was demanding, meticulous work.

The computers were (mostly) good at what they did, but errors had the potential to lead to large losses of life, and a good deal of careful checking was needed to make sure no disastrous mistakes found their way into the printed tables. So a 'comparer' kept an eye on things, standing between Maskelyne and the computers. And here Hutton got involved. It was unglamorous work involving liaison with the computers as well as with Maskelyne. And it was a deeply picky process. The computers worked in pairs, without communicating with each other; one found the moon's position for every midnight and one its position for every noon. The comparer merged the tables and checked that the moon's predicted motion contained none of the implausible jumps that would signal a mistake in someone's calculations. If there was a problem, the comparer redid the calculations himself until everything was right and he had a full month's table of lunar positions for both noon and midnight.

Then he selected some stars that lay close to the moon's path, and sent the complete, correct table back to the computers so they could both, independently, compute tables of the moon's predicted distances from those stars through the month. When the comparer received this information, he checked that the tables drawn up by the two computers were identical and, once again, sorted out any discrepancies by repeating the work himself if necessary.

He also prepared various other pages of the *Nautical Almanac* such as the initial explanation of symbols and a chart showing the positions of Jupiter's satellites. And finally, when the almanac was being printed, he corrected the proof sheets: yet more checking of long tables of numbers that were supposed to be identical.

On and off during 1777–9 Hutton did all this, covering the comparing work for a total of twelve months' worth of *Nautical Almanacs*; he also performed some extra tasks such as checking the predictions for eclipses of Jupiter's moons that were printed in one almanac. He was paid (a total of about seventy-five pounds), but the money was far from being the point. 'Comparer' was a position of significant trust, and Maskelyne did not give it to just anyone. Hutton was very possibly doing Maskelyne a favour by filling in for months when no other comparer had been found or was available. And by doing so, and doing it well, he significantly increased his credit in the network around the Astronomer Royal. He was establishing himself as part of Maskelyne's mathematical/astronomical circle, and confirming his valuable relationship with the Astronomer Royal.

∞

There was more. His work for Maskelyne gave Hutton the opportunity to make contact with the Board of Longitude itself, and to do more work for it on an occasional basis. Through 1779 he corrected proofs of mathematical publications for the board – at a guinea a sheet – and for one book he was paid to translate a preface from Latin into English. In 1781–2 he provided lunar computations apparently outside the normal cycle of work on the *Nautical Almanac*, for which he was paid ten pounds ten shillings 'for my Trouble'. And by 1780 he was writing to the Board to present a work of his own: a book of mathematical tables.

Mathematical tables had long been one of Hutton's interests. His very first book, the *School-master's Guide*, had ended with a little table of the first twelve powers of each of the nine digits, and the 1770 *Mensuration* similarly closed with a thirty-page table of the areas of segments of a circle. During the 1760s and 1770s he had found himself repeatedly making computations involving roots and reciprocals of numbers,

> and as it seemed probable that this might be the case with me for many years longer, I formed the resolution of preserving

all such roots and reciprocals as I should occasionally produce in my calculations, that I might have them always ready on any future occasion; which I did by entering them always in a little book, ruled for the purpose, till I have at last collected to the number of 1000.

He published the resulting table in the 1775 *Miscellanea Mathematica*.

His next venture of the kind was similar, but larger: a stand-alone table of the products and powers of numbers: products up to 1000 times 100; powers up to 100^{10}. It could well have been collected together over a period of years like the table of roots and reciprocals, and when it was complete it enabled the rapid looking-up of over a hundred thousand different products or powers.

The eighteenth century was a period of heroic manual calculation (as the *Nautical Almanac* illustrated), and well-judged, accurate printed tables were of real use to those involved in such work. Hutton rightly saw that the genre offered a route to bring himself once more to public notice. The Board of Longitude accepted his project and agreed to grant him two hundred pounds to see it through the press: nearly as much as his annual salary at Woolwich. Five hundred copies were printed, and as well as being sold they were adopted as a standard part of the *Nautical Almanac* computers' equipment.

The table of powers showed once again that Hutton loved rapid, accurate calculation and was extraordinarily good at it. The preface displayed a virtuosic facility at getting the tables to do more than they at first sight seemed capable of: using a table of squares to compute square roots, using interpolation or repeated calculation to multiply numbers larger than those allowed for in the tables. One reviewer dryly remarked that 'but little trouble will be saved' by using the tables in such complex cases, but that wasn't entirely the point: as before, Hutton was telling the reader something important about who he was and what kind of mind he had.

There was more (with Charles Hutton there was always more). One of the natural goals for a mathematician interested in tables was to work on tables of logarithms. Judiciously deployed, logarithms speeded many types of calculation, and tables of them to six or eight or even more decimal places had been in print since early in the seventeenth century. Computing them was laborious, and getting them right in detail – and printed correctly in detail – was famously hard; the standard work, *Sherwin's Tables*, was notorious for its inaccuracy by the time of the 1771 fifth edition. Hutton himself had compiled a list of several thousand errors in that book, and he reckoned the time was right to begin again from scratch.

He conceived a project to print new, freshly calculated logarithm tables to seven places, together with supplementary smaller tables, enabling the user to find the logarithm or antilogarithm of any number to twenty or to sixty-one places. Through the early 1780s he worked on these: they represented thousands of hours of work on top of everything else he was doing. Calculate, recalculate. Check the calculation. Transcribe into a fair copy. Check the transcription. Quiet, endless scratch of the quill against a background of ordnance testing, cadets drilling (or rioting) and military bands practising. You wonder how he found the discipline, the energy, or even just the time.

In fact, he didn't, or not all of it. The manuscript of his logarithm tables survives, and it clearly shows he got his family to help him. The rough work and the fair copies are in at least two different hands: Margaret's almost certainly, and quite possibly those of one or more of his older children. A biography that appeared later during Hutton's lifetime, indeed, acknowledged that the labour and calculation for the product tables was 'chiefly owing to the industry' of Margaret, who assisted him with other, unspecified laborious calculations too.

It wasn't an unusual arrangement for women to work as unpaid assistants in support of their husbands' work. Acknowledgement was rare, so we don't know exactly how common it was. Margaret's hand also possibly appears making comments and revisions in some of Hutton's scientific manuscripts from the 1780s. His daughter Isabella would later become Hutton's main amanuensis; when old age made his own handwriting wobbly she wrote nearly all his letters for him. The youngest, Charlotte, was not as yet old enough to be involved, but her turn would come.

So, rather than heroic solitary labour on the tables – or perhaps on anything else – we should imagine a sort of bustling family workshop, in which drafts were passed from hand to hand and might receive work from two or three different people. The name of Charles Hutton alone appeared on the title pages. But others played a role, as he occasionally acknowledged, and without them he would almost certainly have been unable to complete the volume of work he did.

The point perhaps deserves pressing slightly. Within the philomath world, women were visible, though they were not as numerous as men. In *The Ladies' Diary* some women hid behind male or neutral pseudonyms; only a few appeared under their own names. In the world of the *Nautical Almanac*, too, one woman was involved as a calculator in this period. Male-dominated but not male-exclusive worlds, then, as far as the evidence goes. The glimpse Hutton's manuscripts provide of the participation of his family suggests that women's roles may have been more substantial than we think, more than we can usually see. That, when we read a publication by 'Charles Hutton', we're not hearing his individual voice alone but a composite, in which other voices from his household, both male and female, are involved.

Meanwhile, the calculations were done. Hutton added an enormous preface, setting out the history of logarithms and their calculation over the previous two hundred years and detailing how his own tables had been calculated and how to use them. More than anything he had done so far, the preface attempted to establish him as more than a mere technician; someone who was not only good at mathematics but knowledgeable about the subject in a humane, humanistic way. A vast range of reading was displayed; information from two centuries was synthesised and formed into a narrative. Judgements were passed, credit was assigned or reassigned. Hutton, here, was reinventing himself as an authority on the history of mathematics, on its nature; and perhaps even an authority on where mathematics should go next, what it was useful for, why it mattered.

There were delays in the calculating work, for reasons we shall hear about in Chapter 6, and there were delays in printing the book, when Hutton insisted on a demanding programme of checking that entailed comparing each proof sheet with the manuscript several times. Almost certainly, other members of his household were involved once again. He was evidently confident of their work; and in a printed list of errata he admitted to only seven mistakes in the logarithm tables when they finally appeared.

The tables were admired by reviewers; it was hard to see how one could do much else, since time alone would determine whether they were really as accurate as Hutton said. On the whole it seems they were, and they remained in print until 1894, through thirteen editions. Like Hutton's table of powers they became part of the standard set of books loaned by Maskelyne to *Nautical Almanac* computers and comparers. Reviewers noted the immense erudition of the preface and the authoritative style:

a judicious arrangement of the subjects treated of, and a clear unambiguous statement of propositions and their demonstrations, are the leading excellencies in subjects of this kind; and these, we will venture to affirm, are no where more conspicuous than in the work before us.

∞

At the same time, in a different mathematical world from that of the Royal Observatory and the Board of Longitude – but certainly an overlapping one – Hutton was also networking; working his earlier connection with the world of the Georgian philomaths and *The Ladies' Diary* in particular. His collection of material from the *Diary* had continued to appear in separate numbers across the period of his move to London, and when it was collected together in volume form in 1775 it received some favourable notices in the London papers. The *Diary* itself continued to appear each year, of course, but during 1773 its editor, Edward Rollinson, died. There was certainly no more natural choice to replace him than Charles Hutton, and within days of his arrival in Woolwich, later that summer, he was approached by the Stationers' Company of London – which published the *Diary* – and asked to do so. By 1775 he was established as editor of the *Diary*, and he held a sale of Rollinson's books from his house in Woolwich. In 1781, following the death of another editor, he signed an agreement with the Stationers' Company to compile not just the *Diary* but several other annual almanacs too; he was in charge, at his peak, of at least nine different titles, although a few remained in other hands.

The *Diary* work chiefly involved dealing with the substantial numbers of letters it received, containing solutions to the previous year's problems as well as proposing new problems. Most arrived in a rush before the deadline of 1 May. It was not just mathematics: there were also riddles in verse, questions on various topics. Regular notes from the editor to keep contributions short, and in some years the drastic expedient of printing some parts in almost unreadably small type, indicate the flood of material that was coming in. At times Hutton kept material back for two years before he could find space for it.

There was sifting of solutions, checking their correctness and choosing which were well enough expressed to be printed; there was judgement of the literary merit of the poetic compositions and questions. There was the occasional act of censorship; in 1775 Hutton noted that 'The enigma on a Candlestick is too indelicate for insertion.' There were complex queries from contributors to deal with, and accusations of plagiarism or false attribution. There were proof sheets to check.

∞

If all this gives the impression of an immense and perhaps an uncontrolled busyness, it should. Hutton was developing quite a habit of taking on mathematical odd jobs. For Maskelyne and to some degree for the Board of Longitude he was becoming a preferred choice for computation, mathematical proof-reading, and even mathematical translation work. With his almanac work and the new editions of his early books (there were at least five during the decade from 1775 to 1785), it begins to be surprising he had time for his teaching. He also reviewed mathematical books – anonymously – for the monthly periodicals. His letters comment on how busy he was, and his prefaces on the degree of interruption some of his projects received from others.

It was relief from teaching elementary geometry to unruly cadets, and it was visibility and status of a kind. But much of it left him rather out of sight, in danger of seeming a mere technician. Computers and comparers were not credited in the *Nautical Almanac*; the editor of *The Ladies' Diary* was well paid but nowhere named in its pages. The production of a book of mathematical tables made one a 'compiler' or an 'editor', not an 'author'. But Hutton also aspired to a higher level of visibility and status (and reward). There were wider worlds than that of the Board of Longitude and more prestigious ones than that of the philomaths.

Indeed, as early as March 1774, less than a year after his arrival in Woolwich, Hutton had been proposed for fellowship of the Royal Society, the premier scientific society in Britain. Colonel J. Phipps, who plays no recorded part in Hutton's life besides this, seems to have been the moving force, though ten other Fellows signed the proposal. They inevitably included Nevil Maskelyne as well as Samuel Horsley, another of the examiners for the Woolwich job; also named was the influential botanist Joseph Banks, one of the rising men at the Society. The certificate stated that Hutton was 'well acquainted with Mathematical and Philosophical Literature', a claim with which few would have argued, 'a Gentleman' and 'likely to become an useful Member'. After a lumbering process, involving delays during the summer recess, he was elected in June and formally admitted to fellowship on 10 November.

Hutton used his status as FRS when advertising his books, and he usually attended one meeting of the Society out of two: as many as he felt he could make time for. He published a few papers in the Society's *Philosophical Transactions*, though these were modest affairs: a brief solution of geometrical problems; a six-page suggestion to construct trigonometrical tables using a new set of units (radians rather than degrees); a new algebraic trick for finding infinite series whose value was equal to that of π; a lengthy, pedestrian article surveying the known methods of solving cubic equations. Attention-grabbing these weren't, and they remained firmly within the sphere of technical mathematical work rather than front-rank natural philosophy.

But Nevil Maskelyne had a specific project in hand which would ultimately shed glory on both himself, his country, and his friend Charles Hutton. In 1772 he had proposed to the Royal Society that it use some money left over from another project to undertake experiments on the gravitational attraction of mountains. Part of the point was to demonstrate that there *was* such an attraction; a

French experiment in the 1730s had been inconclusive, and it would be an important confirmation of the Newtonian theory, which stated that every object in the universe exerted a gravitational attraction on every other, proportional to its mass. The experiment would 'make the universal gravitation of matter palpable', as Maskelyne put it. It would also, if the experiment was done carefully, enable the experimenters to work out the *strength* of the gravitational attraction of a given mountain, and work back from that to deduce the mass of the earth, a quantity that was so far a matter of mere vague guesses.

How to do it? Simple enough: find a reasonably isolated mountain and set up a plumb line next to it. By comparing the line's direction with the background of stars, you can see how far the mountain is deflecting it from vertical. Delicate work, but quite feasible with eighteenth-century equipment.

Charles Mason (one of the men who measured the Dixon Line) worked on the selection of a site in 1773; he considered several mountains and chose one in central Scotland. With a mile-long ridge running east–west, and fairly isolated from other hills, bleak Schiehallion he reckoned was as close to ideal as the experiment was going to get. Reuben Burrow, once an assistant at the Royal Observatory and an acquaintance of Hutton, went north and surveyed the site with a local man named Menzies in 1774 and 1775, and after a certain amount of foot-dragging Maskelyne himself went up to run the astronomical observations. They lived in tents and huts and suffered greatly from bad weather (it was rumoured that the name *Schiehallion* meant 'constant storm' in Gaelic). But eventually the work was done: 337 observations of 43 different stars from two different locations. When Maskelyne worked through the numbers, it turned out that the plumb lines in the two places, on either side of the mountain, really did point in slightly different directions, beyond what would be expected from their difference in latitude.

This was success insofar as it showed that the hill exerted a gravitational attraction, and the divergence of the two plumb lines (three thousandths of a degree or so) enabled Maskelyne to deduce that the density of the earth was about double that of the mountain. Read to the Royal Society in July 1775 and published in the *Philosophical Transactions* later that year, his paper won Maskelyne the Society's prestigious Copley Medal. As he put it, everyone was now obliged to be a Newtonian, at least as far as believing in the universal mutual attraction of matter.

But there was more work to be done. Detailed processing of the survey data could potentially yield a much more accurate estimate of the density of the earth. It would require a mass of calculation, determining the shape of the mountain from the survey and then working out precisely its expected gravitational effect at the two observation points.

Maskelyne himself declined to do it: too laborious. It can't have taken him long to think of a suitable man for the job; indeed, he had already discussed with Hutton some of the computational techniques he might use. The Royal Society agreed to pay Hutton, and even met his rather cheeky request for a new set of quality drawing instruments to help him make plans and diagrams.

For about a year in 1777–8 Hutton devoted much of his working time to the project; an 'immense labour'. The easy part was taking the survey data and using it to prepare a detailed map of the site. As far as we know this was only the second or third actual map Hutton had prepared, but the mathematical techniques were well known; indeed, they were detailed in his own textbook on mensuration. The quantity of work, nevertheless, was huge. The surveyors had taken more than seventy sections through the Schiehallion site: most vertical, some horizontal and some neither. In all they described nearly a thousand points, each fixed in position by taking angles from other points.

Several thousand trigonometrical calculations were needed to turn the data into a map four feet square covering the mountain and its surroundings.

To one of his versions of this map Hutton added contour lines, 'connecting together by a faint line all the points which were of the same relative altitude'. Although people had described the idea before, Hutton seems to have been the first person actually to make a contour map, and it's a pity the map itself has not survived.

The really hard part came next: given the contour map of the mountain, to work out the expected gravitational attraction of the hill at each of the two observation sites. After at least one false start, and with the help of some ideas from fellow FRS Henry Cavendish, Hutton came up with a feasible way of doing it. Instead of dividing the map into a grid of squares, he divided it up using circles and parallel lines. For each of the two astronomical observation points Maskelyne had used, he drew twenty rings, crossed by twelve parallel lines in each quadrant, making a total of nearly a thousand irregularly shaped patches of land. Or rather, nearly a thousand irregularly shaped pillars of rock exerting their separate, determinable gravitational effects on the observatory. Adding up all the gravitational effects involved a specially constructed slide rule and a computational trick suggested by Cavendish.

The end of it all was that the hill should be expected to attract the plumb line just under ten thousand times less than the earth did. The astronomical observations showed that in fact it attracted it nearly eighteen thousand times less, so Hutton concluded that the hill was less dense than the earth by a ratio of eighteen to ten, or nine to five. To put it another way, the earth, taken as a whole, was 80 per cent more dense than Schiehallion. Supposing the mountain was made of common rock weighing 2,500 kilograms per cubic metre, it followed that the earth, taken as a whole, had a mass of about 4,500 kilograms per cubic metre of its volume. (Kilograms were not yet invented; Hutton expressed everything in relation to the density of water, which comes to much the same thing.)

Hutton's paper was a virtuoso performance: certainly the most spectacular piece of work he had yet done. To his death 'weighing the earth' remained one of his proudest achievements. In the *Philosophical Transactions* the paper filled a hundred pages, with tables of data, example computations, maps, endless details about the site and the mathematics. As an additional reward for those who made it to the end, he deduced the densities of the other planets from that of the earth (the Newtonian model of how things move in the solar system makes it possible to do this) and speculated about the internal structure of the earth, inferring the existence of 'great quantities of metals, or such like dense matter' deep inside the planet. Roughly half the matter in the earth would need to be metal to account for his proposed overall density.

It was a piece of work that sat very neatly on the boundary between the worlds of the technician or assistant and that of the original natural philosopher: facing both ways, or rather displaying two faces of its author. Once again, Hutton presented himself as someone who relished calculation for its own sake, and who loved details; the paper described minuscule corrections for the effect of temperature on the surveyors' measuring rods and for the fact that by the end of the work they had worn to a shorter length than they started; it described rubbing out certain lines on the plan in order to leave other information clear. The work was in a clear sense firmly under Maskelyne's protection, and Hutton had been directly paid by the Royal Society to do it. But at the same time, he also presented himself as a natural philosopher in his own right: someone who was thoroughly in control of the large picture, the overall scheme that made the individual calculations mean something. Someone who could turn a book of numbers into important information about the earth and the planets. Against the background of the Royal Military Academy and its wartime chaos, Hutton had flown in thought to every planet that orbited the sun, deducing their densities and laying the foundation for deductions about their structure. He had travelled to the interior of the earth and seen in his mind's eye the metals that lay there. And he had not only felt the force of universal gravitation but determined its strength in absolute terms: the first to do so in the century since Newton's discovery.

∞

And still this was not all. Hutton wished to do more to establish a reputation as a working experimentalist, and he was never a man to rest on his laurels. During summer 1775 he devised a programme of experiments in ballistics that he would carry out at Woolwich, ably assisted by the men and cadets of the Royal Artillery. They took place under the auspices of the Woolwich Military Society, set up in 1772 to promote military science and experiment. Hutton was the society's secretary for a time. Military oversight came from Thomas Blomefield, captain-lieutenant in the Royal Artillery and aide-de-camp to the Master-General of the Ordnance; but it was Hutton who both determined the direction of the ballistics experiments and, usually, oversaw them from day to day.

He was far from being the first in the field; there was a long history of trying to describe the flight of projectiles using mathematics. But it was only in the last thirty years or so that experimenters had found ways to look empirically at what happened when a ball came out of a gun. How could you possibly measure such rapid motion?

The clever answer had been found by Benjamin Robins, a mid-century mathematical practitioner and engineer who missed out on a job at Woolwich in his day. You fire your gun at a large pendulum, and you measure how far the pendulum swings as a

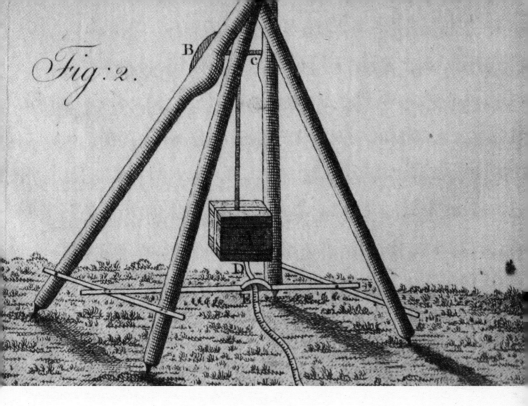

Hutton's ballistic pendulum.

result of being hit. Knowing the weight of the ball, the weight of the pendulum and something about the behaviour of pendulums, you can work out the ball's speed. Robins used the so-called ballistic pendulum to revolutionise the study of artillery – his device has been called 'as much a scientific milestone as Galileo's telescope or Watt's steam engine'. 'There can have been few cases in history of one man changing the state of knowledge of any branch of science by so great a step.' But he only built fairly small pendulums and used them to study musketry; the subject badly needed work on a larger scale.

So Hutton built himself a really big ballistic pendulum.

It was about fifteen feet high, and the bob used in the first series of experiments weighed 328 pounds. A brass cannon was procured, and supplies of powder and shot. Over the summer of 1775 and the following two summers, teams of gunners under Hutton's direction fired balls at the pendulum, varying the size of the powder charge in order to find how it affected the speed of the shot. The explosive properties of gunpowder were in some respects not well understood. Should the ball fit the cannon exactly, or should there be a gap: 'windage', so called? Too tight a fit and artillery pieces would be liable to explode. Too much of a gap and some of the explosive force would be lost. Too much powder and it wouldn't all burn (unburnt powder frequently blew out of the mouth of Hutton's cannon, scattering the area and even the bystanders).

It was once again laborious, meticulous work, but of a very different kind from the meticulous calculation that had so far been the foundation of Hutton's success. In addition to the regular drill of loading and firing a cannon, it was necessary to size and weigh each ball before firing; after firing, to check and record the pendulum's recoil; and, if there was a hole in the pendulum, to plug it and adjust the weight of the pendulum in the subsequent calculations. Pendulums regularly had to be repaired, rehung and reweighed and their centres of gravity redetermined. Different sizes of charge had to be made up and kept ready for use; thermometer readings were taken on each day of work. At first, the team managed only about four shots per hour.

If it hadn't the briskness of a pitched battle, it had at least a hint of the dangers. The pendulum was understandably apt to fall to pieces under bombardment, and Hutton's notes regularly report that shot passed through it, pieces fell off it, its metal bands burst. One ill-advised attempt to strengthen the pendulum with lead resulted in a cannon ball breaking to pieces when it struck, firing splinters of metal in all directions. In later series of experiments, when the shots were being fired from longer distances, an alarming proportion of them missed the pendulum altogether, burying themselves in the earthwork behind it. Hutton perhaps brought to such efforts some recollection of the physically tough environment of his youth, and the men and cadets of the Royal Artillery were of course no strangers to flying metal. Hutton was lucky no one was ever hurt; there's no record that any of those involved in the experiments had any kind of cover.

It was hard to get it all right. Gunpowder varied in quality; gunners rammed it with varying degrees of vigour. The measuring tape attached to the pendulum flapped around in the wind and became tangled. Even when all the equipment behaved properly, all Hutton could do was record the displacement of the pendulum for later calculation; working out the actual velocities of the balls was not something to be done on the spot. Only later, in his study, did Hutton find out just how fast the balls had been moving or how their speed varied from one shot to the next.

By the end of three summers, experimenting as and when opportunity offered, Hutton had fired several dozen rounds of ammunition into giant pendulums, and he was in a position to write up what he had found so far. His conclusions were several. A cannonball's speed varied as the inverse square root of its weight: quadruple the weight and, other things being equal, the ball would go half as fast. The speed also varied as the square root of the charge: quadruple the charge and the ball would go twice as fast. He noticed that decreasing the windage greatly increased the ball's speed, and recommended strongly that this be pursued in practice.

There was a first account of all this for the Military Society and Lord Townshend, the new Master-General of the Ordnance. Hutton struggled with this piece of writing, his first attempt at the role of natural philosopher, and Margaret read over his drafts and improved his prose in places. The work had been a good deal of trouble and had cost the Board of Ordnance sixty-odd pounds, and Hutton was determined that it should seem to be of good value and the work be authorised to continue. He wanted to study air resistance, how shots' range depended on their speed, and how different kinds of gunpowder affected matters.

Townshend did approve, and Hutton went on to write up the work for presentation to the Royal Society. It was by far his most ambitious paper yet, intended to establish his reputation as not just a backroom calculator but a front-rank natural philosopher.

The paper was read to the Royal Society on 8 January 1778 and printed in the *Philosophical Transactions* later that year. As with some of Hutton's other performances, the author's name concealed the work of quite a number of other people in procuring equipment, building and repairing pendulums, firing cannon and recording data. Margaret's revisions of the text received no credit; neither did the accurate drawings, probably done by one of the more talented Woolwich cadets, that accompanied the paper in print.

The military importance of the work spoke for itself, but Hutton stressed in his paper that his results mattered not just for the accurate firing of cannon but for all 'those parts of natural philosophy which are dependent on the effects of fired gunpowder'; for the study of explosions planned and unplanned, of the composition of the air, of the behaviour of gases. And the Royal Society was collectively convinced. By July the Society's Council had voted to award Hutton the Copley Medal for his paper on ballistics.

Hutton's toil had paid off. The Copley was the Society's only medal for intellectual distinction: not just the highest but almost the only scientific honour in England at this time. The ceremony of its presentation took place on the Society's most solemn day of the year, the anniversary of its foundation on St Andrew's Day in 1662. In his speech, the president of the Royal Society, John Pringle, mentioned that the Society made the award 'with the more cordial affection, as by your other ingenious and valuable communications they are assured, not only of your talents, but of your zeal, for promoting the interests and honour of their Institution'.

For a few months Hutton was quite the golden boy at the Royal Society. Samuel Horsley was stepping down as secretary of the Society in the same year, and Hutton was given to understand the job was his if he wanted it (he did). While writing his speech Pringle got to know him and found him a congenial man; he invited Hutton to the prestigious and exclusive 'Club of the Royal Philosophers'. Hutton described the scene with relish:

> there was a select number of gentlemen who constituted themselves into a certain club, that, after the usual weekly meeting of the Royal Society, retired to Slaughter's Coffee-house, at the top of St. Martin's Lane, to eat a few oysters, and hold familiar discourse together on the subjects that had occurred at the Society's meeting, and on any other current scientific matters.

This was Charles Hutton's apogee. A few transcendently busy years, and a great deal of networking, had transformed him from provincial mathematics teacher to the darling of the Royal Society; brought him out of the shadowy world of backroom calculation to the sunlight of public admiration. From the secretaryship of the Royal Society he could perhaps dream of a further step to vice-president or – who knew? – even the presidency itself.

6

Foreign Secretary

8 January 1784. Somerset House, on the Strand in London. A meeting of the Royal Society. A big panelled room that echoes when you talk; nearly 170 men seated in rows. The scene dominated by the president's chair, his table, his ceremonial mace.

For all the finery – the wigs, the lace, the gilt – it's no genteel scene. Samuel Horsley, bishop and mathematician, is making a speech, or trying to. He's been on his feet for some time, and he's getting hoarse from shouting. People are clattering their sticks, shouting him down, clamouring for him to come to the point. Noise, confusion. The president isn't in control, isn't trying to be in control.

Horsley, goaded beyond civility, cries at last that he and his friends 'can at least SECEDE . . . the President will be left, with his train of feeble *Amateurs*, and' – pointing to the mace – 'that Toy upon the table, the GHOST of that Society in which Philosophy once reigned and Newton presided as her minister.'

In the uproar that follows Nevil Maskelyne, Astronomer Royal, jumps to his feet and roars, 'Yes, sir: for where the Learning is, there will the Real Royal Society be.'

∞

At the meeting in 1778, at which Hutton was presented with the Copley Medal of the Royal Society, and a lengthy speech made in his praise and honour, his success in that particular world also showed the first, small signs of unravelling. There was an upset about the secretaryship of the Society that Hutton had been promised; a second candidate had declared himself, campaigned vigorously, and on a vote the role went to him. This was Paul Henry Maty, a clergyman who had recently developed career-ending scruples (had discovered he didn't believe in the Holy Trinity, to be exact). Hutton and his friends resented what had happened, and felt Maty had lobbied harder than was decent. But there was no denying that his need for employment was great. Hutton was consoled with election as an ordinary member of Council and promised he could have the *next* secretaryship that fell vacant; and after a little dithering he was also given Maty's previous position as foreign secretary to the Society.

Though foreign secretary may sound similar to secretary, it was not. The two ordinary secretaries read out the papers at the weekly meetings, and were highly visible and influential. The foreign secretary occupied a backroom position of exactly the kind Hutton wished to leave behind. The written instructions saw the role as comprising two parts. First, the foreign correspondence proper: answering letters to the Society from overseas; letting foreign members know that they had become foreign members; acknowledging their gifts, usually of books. In practice the gifts were much the most frequent occurrence. A Council decision of about a decade earlier specified that thanks should be returned using a pre-printed form, to minimise any impression that judgements of merit were being passed on the books. Not, then, a position of much responsibility or intellectual stimulation.

The second, rather more interesting, part of the foreign secretary's role was to translate foreign papers – usually from French or Latin – for the Society, so that they could be read at a meeting and, if desired, printed in the *Philosophical Transactions*. Practice varied somewhat; for papers in modern languages the *Transactions* carried both the original and an English translation, while for Latin papers the original was printed and the translation merely kept on file for those who wished to admit they couldn't read Latin and consult it.

Hutton read French – and Italian – fairly confidently, and he had already carried out paid translation work from Latin for the Board of Longitude. He was sure he could acquit himself well in the role, and so he did. If the translating was bulkier than the correspondence, it was by no means onerous. Latin papers in the *Transactions* during the three years Hutton was in the role numbered five; French four, Italian two and Swedish one. Presumably he called on outside help for the Swedish.

More varied and perhaps more interesting were his duties as a member of the Society's Council. It met every month for most of the year, and Hutton attended diligently. Council made decisions about the day-to-day running of the Society, as well as selecting from the papers submitted by Fellows those to be read at meetings and printed in the *Philosophical Transactions*. Occasionally papers were referred specifically to Hutton for judgement. Council also oversaw the running of the Royal Observatory and carried out an annual visit to that institution, in which Hutton took part.

From a distance it certainly looked as if all was well, despite the upset about the secretaryship, and as if Hutton was now firmly embedded in the most elite scientific circle Georgian Britain had to offer: busy, influential, prestigious. To add to his prestige, in the summer of 1779 two of Hutton's friends, joint professors of mathematics at Edinburgh, arranged for him to receive an honorary doctorate from that university; Professor Hutton was now (also) Doctor Hutton.

Yet all was *not* well. At the same meeting at which Hutton had missed out on becoming secretary of the Royal Society, his new friend and supporter Sir John Pringle stepped down as president.

Sir Joseph Banks.

He was replaced by Joseph Banks, a man with rather different interests and priorities. Over the next couple of years he would become Hutton's decided enemy.

Had there been a tiff? a slight? a disagreement? (Just a few years before, Banks had signed Hutton's nomination for fellowship.) The mores of the time hindered close scrutiny of the affront, if such there had been. Certainly the cold official minutes of the Council show Hutton behaving neither unusually nor offensively towards the president. Some of the best-informed of contemporaries denied that there had been any specific incident at all, and that appears to be correct.

On one level, though, there was certainly a clash of personalities. Joseph Banks had great qualities, but he liked his subordinates docile, and that Charles Hutton was not. It was his habit to express his views 'with freedom and firmness' at Council, and he may simply have succeeded in irritating the new president beyond endurance. Banks would later describe Hutton as 'a forward young man'; perhaps he was too forward for Banks's liking, and turned out to be someone Banks could not work with.

There were several other factors. One was lingering doubts about Hutton's presentability. He retained a northern accent, and would do so to the end of his life, and – although he did nothing to advertise his pit-boy origins – some could detect from his manner that his gentlemanly status was quite new. One journalist made a joke of the claim that Charles Hutton the Newcastle schoolteacher, author of the *School-master's Guide*, the *Mensuration* and the *Principles of Bridges*, 'could not possibly' be the same man as Professor Hutton, Fellow of the Royal Society and lately awarded the Copley Medal. A crueller later critic, also writing anonymously, found it difficult to imagine how a man of Hutton's background could ever have functioned as foreign secretary of the Royal Society without constantly giving offence.

Hutton was moreover strongly identified with mathematics; and mathematics was something of a cross for the Royal Society to bear. Its weekly meetings consisted of the two secretaries reading out papers sent in by Fellows: a sort of literary re-creation of experimental or observational work done far away. When the subject was the birds and beasts of Botany Bay, rocks and fossils in northern Britain, or even spectacular new experiments on hydrogen gas (or gunpowder), it worked very well.

But mathematics never lent itself to that kind of performance. If a paper was dense with algebra, if it needed its geometrical diagrams in order to be understood, or if – worse yet – it was actually about mathematics rather than its practical uses, there were real difficulties. Even the most skilful reader could scarcely make such things readily comprehensible, and not all Fellows were remotely interested in any case. Some Fellows persisted in sending mathematical papers to the Royal Society, but it was increasingly felt that by doing so they were causing a problem, breaking the rules of a world that was still to some degree about 'polite' discussion of accessible topics.

And Banks made no secret of his personal dislike for the subject. In public he used what a follower of Hutton's called 'waspish and petulant expressions' whenever mathematical papers were read; in private he reckoned mathematics a mere tool, and one whose injudicious use obscured matters. Of a book about electricity he complained that the author had 'done little but apply Conic Sections infinite series & Fluxions to explain the laws of Electricity which I look upon in the same light as driving it like a Fox into an Earth from whence our electricians will never be able to dig it'.

If all that were not enough to damn Hutton, he was also, of course, strongly connected with Banks's predecessor Sir John Pringle. Naturally Banks wanted to consolidate his own power at the Royal Society by replacing Pringle's men with his own wherever it was decent to do so. Banks's own were most usually aristocrats (Banks himself became a baronet in 1781) and gentry, and had fashionable interests such as natural history, horticulture, agriculture, antiquarianism. He was a skilled manager of those constituencies, and he could little afford places on the Council of the Royal Society for someone who was a member of none of them.

Furthermore, Pringle's circle had become politically suspect. During the 1770s Pringle had vocally supported Benjamin Franklin, then in England, in a dispute about the best shape for lightning conductors; it had become a public issue when the Board of Ordnance asked the Royal Society's advice on the subject, needing to protect its gunpowder stores from lightning damage. Pringle and Franklin were opposed by a number of influential Fellows of the Royal Society, who thought their advice wrong and dangerous. For political reasons, Franklin had ceased to be an acceptable figure in Britain by the end of the decade, and it was rumoured Pringle's resignation as president had been called for by the King as a result. Hutton was unavoidably tarred with this brush; many of his own friends were radicals, nonconformists or both, and to make his whiff of radicalism worse he certainly had some personal sympathy for the cause of American liberty.

So things were against Hutton from the start, and although the detailed balance between these factors is now lost to us, it's certain that by the end of two years on Council Banks had had enough of him. It became clear, indeed, that he felt Hutton should never have been elected to fellowship. At the end of 1780 Hutton's name was omitted from the house slate for the new Council. Nearly half the twenty-one or so members of Council were new each year, so his two-year stint was not exceptionally short. But it was a clear signal that he no longer enjoyed the kind of favour that had brought him into the Society, its dining club and its Council during the 1770s.

Banks expressed his irritation in some ways that seem rather puerile. During 1781 the Royal Society put together a printed list of its members. It decided that members' titles would be included, but in August the Council intervened specifically to rule that Hutton should not be styled 'professor' in the list. No other Fellow was singled out for comparable attention. Hutton spotted the clumsy slight when the list was printed and – perhaps unwisely – fought it, producing his warrant from Woolwich and obtaining an under-taking to style him 'professor' in future lists.

Banks's next move was to interfere with the terms of the foreign secretaryship. Early in 1782 he represented to Council that foreign correspondence was not being dealt with adequately, and stated or implied that this was Hutton's fault. After deliberation Council agreed to alter the written instructions for the foreign secretary, striking out the matter of translations and insisting that he should now handle *only* the foreign correspondence proper, such as it was.

Hutton and his friends reckoned this was a demotion, another deliberate slight and perhaps one aimed at provoking him to resign. He had enjoyed the translation work, he was sorry to be deprived of it and reckoned it a waste of money to contract it out as was being proposed. And he had little enthusiasm for answering the Society's (very few) foreign letters. He hesitated, demanding a written copy of his new instructions and stalling for some months before agreeing to continue in his post.

There were admittedly two sides to this. As others saw it, the change was more like a promotion. It took Hutton away from laborious and unglamorous translation work and gave him direct contact with a wide range of distinguished foreigners. Completing and sending pre-printed forms of thanks for books received need not be the whole story. There were occasions when a few well-turned phrases in Latin or French from Hutton himself might be appropriate: might even be necessary, if offence was to be avoided. Such excursions in international *politesse* could have done Hutton himself a great deal of good.

So it's possible this was meant not as a slight but as an olive branch: a chance in Banks's eyes for Hutton to redeem himself, and potentially a valuable gift to an ambitious young networker.

Hutton refused to see it that way. He was up to the task: polite he could do, even obsequious, and if his written French wasn't the most accurate he was reasonably *au fait* with the set phrases of epistolary politeness. Help could easily have been found had he felt he needed it. But instead, he complied stolidly, minimally with his new written instructions, sending the routine forms of thanks and occasionally conferring with the other secretaries about specific items. He used no initiative and penned no personal notes. More complex letters were rare: the records of the Royal Society show that in the 1780s it received perhaps three or four foreign letters a year. Hutton worked on some of them, translating extracts and writing a couple of replies; others were taken away and dealt with by the main secretaries or sometimes by other Fellows who happened to know the correspondents. His performance as foreign secretary attracted no further comment at Council for a year and a half.

He and Banks meanwhile continued in something of a stand-off. Although he was beginning to find his visits to the city tiresome, Hutton continued to attend every second meeting of the Society – keeping his rooms in Clement's Inn in order to do so – and he regularly checked the Society's letters book for anything he needed to deal with. But he and Banks never spoke to each other at the meetings, and Hutton never went to Banks's home in Soho Square: neither to the 'breakfasts' hosted there every Thursday nor to wait on him privately. Banks, for his part, made no overtures either.

Hutton was not the only one who found Joseph Banks a problematic figure at this time. The president was developing a habit of interfering in the election of new Fellows to the Society; on several occasions he declared himself opposed to particular candidates, and he sometimes pressured individuals to blackball candidates of whom he disapproved. His reasons were ultimately to do with the different constituencies that were the bases of his own power, and the various networks of which he rightly or wrongly disapproved. In his world there was room in the Royal Society for both the patrons and the working men of science, but there was none for mathematicians, schoolmasters or commercial authors, bringing with them as they did the taint of interested motives and, often, of radical politics.

Early in 1781, for instance, Banks blocked the election of Henry Clarke, a Manchester mathematics teacher who had the support of both Maskelyne and Hutton. Others he blocked were likewise provincial schoolmasters, country doctors and the like. Some felt Banks was abusing his power, and it began to be said – in a memorable but unfair simplification of his complex set of agendas – that

Banks wished 'to make *great* Men Fellows, instead of *wise* Men'. Maty, as secretary, was particularly troubled and seems to have come in for more than his share of bullying by Banks over this matter of elections.

There were other kinds of conduct from the president that some found overbearing. As well as in elections, he was throwing his weight about in the selection of papers for reading at meetings and printing in the *Transactions*; he got involved in a silly dispute about the reorganisation of the Society's books, over the head of the official librarian, and wasted money – as some thought – on fine new furniture for important visitors and on the framing of pictures. Little things, but they added up to a strong sense among some of the most active members of the Royal Society that Banks was the wrong man for a difficult job.

Ultimately this tangle was not so much the fault of Banks as a consequence of the structure of the Royal Society. It was – by history and by convention – not a learned academy on the European model, but a gentlemen's club. Most of its members were gentlemen with a passive interest in natural philosophy – or at least a vague feeling of goodwill towards it – and most attended few meetings. A minority were actively engaged; and their needs, wishes and views were very different. In a sense, then, he was the president of two different societies, with different memberships and different agendas. The Society being in constant need of both money and social prestige, Banks can hardly be blamed for sometimes favouring rank over intellectual worth, for not always doing what the more active members would have preferred. Yet as president he was also obliged to preside over the weekly meetings, at which the active members preponderated. It was a situation that required a more delicate hand than Banks possessed.

∞

The catastrophe came towards the end of 1783, when the American war was ending and, in Britain, one of the great political upheavals of the century was taking place. Charles Hutton's supporter Sir John Pringle had died the previous year. After years of unconcealed dislike, Banks at last moved decisively against Hutton.

In November, Banks once again told Council there was a problem with Hutton's work as foreign secretary. He didn't say quite what it was, relying – he hoped – on Council's willingness to dismiss Hutton at his word. But Hutton had one friend who was still a member of Council. Nevil Maskelyne protested – as did Paul Henry Maty – and the proposed measure was softened. Instead of sacking him outright, Council ruled that the foreign secretary should be required to reside in London. Since Hutton's position at Woolwich required him to live outside the city, the effect was much the same.

Days later he duly presented his resignation. Not to Council, though, but to a full meeting of the Royal Society.

> Understanding, Sir, that the circumstance of my residence, for a great part of my time, at the distance of nine miles from town, has occasioned, or has been imagined to have occasioned, some difficulty or inconveniences, I therefore beg leave to return thanks for all favours, and to give notice that I wish to resign that office.

It was news to some Fellows that there was such a role as foreign secretary, and news to most that Charles Hutton had been under-performing in it. There was no immediate comment, but Hutton's bearing seems to have struck the room favourably and there was perhaps already a hint of battle in the air.

St Andrew's Day came round, and by old custom the Society voted for its new Council the following day. The house slate was voted in without demur, and Maskelyne was not on it. Excuses could be found – Banks wished him to spend more time on a book he was supposed to be working on – but it took little imagination to think he was being punished for his attempt to support Hutton at Council.

At the next meeting Banks's opponents acted. Edward Poore, an otherwise obscure Fellow of the Society, moved thanks to Hutton for his work as foreign secretary. Seconded by Maty, he carried his motion, though the president and his friends were against it. The incident rattled Banks – as it should have – and the new Council met in a hasty session to reconfirm the decision about where foreign secretaries should live. Banks let slip a little more information about what Hutton was supposed to have done wrong. Certain letters, he said, had been found unanswered, and a protest received from one foreigner who had been sent the mere printed form of thanks.

Hutton now supplied Council with a written defence of his conduct, in which he said he had followed all the written and verbal instructions he had received and pointed out that neither the president nor anyone else had ever drawn his attention to specific letters needing to be answered. It did him no good at Council, but at the next meeting of the Society the mathematician Samuel Horsley successfully moved that it be read out, and the consequence was another motion – passed by a clear majority – that Hutton had 'fully justified himself'. The meeting made no demand that he be reinstated, but the Society had administered an unmistakable slap to its president, and that in front of an unusually large number of guests (fifty-eight), perhaps drawn by the rumour of war. The mood was uncomfortable, Banks's situation excruciating. Had someone proposed a no-confidence motion on the spot, there was every chance Banks could have been unseated as president.

But there was no such decisive action. Instead Horsley took the floor again. Archdeacon of St Albans as well as editor of the works of Newton, he was a celebrated pulpit orator but showed at the

Royal Society an unfortunate tendency to lose control of his manner. On this occasion he kept his accusations against Banks vague. He mentioned presidential interference in elections and in the selection of papers for reading at meetings. He claimed that enough presidential wrongdoing had occurred to keep the Society in discussion all winter. And with that alarming thought, the Royal Society closed for Christmas.

∞

Already the row – since row there was clearly to be – had largely ceased to be about Charles Hutton and the Society's foreign correspondence, and had come to be about other matters, notably Banks's general fitness to continue as president. Most who knew Hutton, indeed, thought him amiable and able, and very few knew or cared about the Society's meagre correspondence with nations with which Britain had until very recently been at war.

The Christmas break saw furious activity. Banks's friends assembled an impressive party of supporters. Banks himself was confident of riding out the storm Horsley had raised; but he overcame an initial distaste for canvassing and held a private meeting of about fifty sympathetic Fellows on New Year's Day. There, further information about Hutton's alleged neglect of specific letters was released; an extract from the letter of complaint, from the Geneva-based naturalist Charles Bonnet, was read out. Many of those present undertook to support two motions Banks's friends would propose at the next full meeting of the Society. To ensure a good turnout from the kind of patrons who were generally absent from the weekly meetings, Banks sent an engraved card to every Fellow requesting attendance since 'it is probable that Questions will be agitated on which the opinion of the society at large ought to be taken'.

Hutton's friends did less. They began a series of letters to two of the London newspapers setting out their side of the story, but the publicity would do them little good in the long run. They had little success in recruiting support from those who didn't normally attend the Society's meetings.

The first of those meetings in 1784 was chaotic. It lasted well over three hours, and saw the rowdiest scenes the Royal Society has ever known. The Society's old meeting room at Somerset House is one of those panelled places that echoes loudly even to footsteps. A good shout raises quite a din, and if several shouted at once the result would be appalling.

The house was visibly packed with Banks's supporters, and no scientific business was done at all; no papers were read. Instead, the Fellows debated a carefully worded motion in support of their president. There were lengthy speeches on both sides. Horsley in particular tried to urge the Fellows to hear the case against Banks before they voted their approval of him. He even tried to name and discuss the candidates Banks had blocked from election. Despite what a later writer called 'a style of eloquence scarcely to be surpassed' he was shouted down, Banks's friends 'riotously clattering with their sticks'. Banks 'took not the least pains to procure silence, and restore order to the debate'. Well-informed observers reckoned Horsley's devotion to the good of the Society was sincere, but his offensive manner may well have done his cause more harm than good. Rumour also said his party aimed not just at ousting Banks but at putting some specific individual in the chair: some said Horsley himself, others Lord Mahon (author, as it happens, of the book on electricity of which Banks disapproved). The hint that personal ambition was involved helped the rebels' cause not a bit.

When it was finally put to the ballot around eleven o'clock, the approval of Banks was carried by a large majority (119 to 42). In a sense all this proved was that Banks had brought in a hundred or so of his supporters for the occasion (and that Horsley hadn't). In a normal meeting the group who had voted against Banks would have been a majority.

Right at the end of the meeting a second motion was carried without contest, Horsley's party presumably feeling that there was nothing to be gained from further shouting in front of an obviously packed house. This was a measure to require two meetings' notice of any motions for discussion in the future. It sounds like a detail, but it was probably Banks's shrewdest move in the whole affair; it would give him two weeks to summon his supporters to any future meeting when his presidency would be threatened.

∞

What followed was rather like aftermath, since the anti-Banks party could now only hope that Banks's supporters would tire of turning up to meetings for the purpose of voting with him. They didn't.

On 29 January a sly motion to print the names of Copley Medal winners in the *Transactions* was passed after another lengthy debate: but without a crucial amendment that would have required a list of past winners also to be printed. Hutton and Maskelyne were past winners; that was the point, and advertising their names and achievements at this moment would have embarrassed Banks and his Council, though it would probably have done nothing worse.

On 12 February Francis Maseres, another mathematician Fellow, brought in a motion to revoke Council's December decision about where the foreign secretary should live, and to ask Hutton to take up the post again. It made for another long, noisy meeting, at which Hutton's written defence was read out once more. Banks was sincerely moved by the situation, but he was confident of success: 'I felt at that moment like a Bull going to be baited', he wrote in a private note, but like a game bull who likes the fight as well as the dogs and 'has more than an equal chance of Success'. Once more he packed the room with supporters, who eventually defeated the motion by 85 to 47.

Finally on 26 February the Society debated two motions whose point was to restrict presidential influence in elections. Yet again a meeting was wasted in shouting; yet again the room was packed with Banks's supporters, making the outcome a foregone conclusion when the vote at last came, around eleven at night.

In the course of four rowdy meetings several members of the anti-Banks party had disgraced themselves quite thoroughly. Horsley repeatedly lost his temper and apparently made something of a spectacle of himself:

> The manner which he assumed . . . will not easily be forgotten. The impression will long remain . . . of the power of voice, and the energy of words, with which his denunciations were delivered. The high tone he adopted went beyond the usual custom of public debates.

He had openly threatened secession at the 8 January meeting, and Maskelyne replied to say that 'where the Learning is there will the Real Royal Society be'. The split in the Royal Society was also to some degree a split by subject matter, natural history being on the whole somewhat more open to the gentleman dilettante than were mathematics or astronomy by this period. Thus the anti-Banks party was composed largely, though by no means exclusively, of astronomers and mathematicians. And Banks understood Horsley and Maskelyne to be referring specifically to the mathematicians. His private response was that 'howsoever respectable mathematicks as a science may be it by no means can pretend to monopolise the praise due to Learning[;] it is indeed little more than a tool with which other sciences are hewd into form'.

Margaret Ord also took up the pen in private, creating in imagination a scene she can only have heard about at second hand. We otherwise have little of her authentic voice, and her flavoursome verses deserve quoting in full:

Tis Horsley's voice loud strikes the ear,
And forceful strikes the guilty chair;
   Agast Sir Joseph stares:
And, husht, around the listning throng,
Nor breathe, nor stir, nor move the tongue,
   While painful truths he hears.

Each member feels the generous spirit;
Indignant glows for Hutton's merit;
   In virtue's praise all join:
The gods themselves espouse the cause,
And Pringle's ghost with warm applause
   Approves the sounds divine.

Go on Horsley! in Newton's plan;
Fair truth directs thee, generous man!
   To save celestial science.
Good men and true the steady hand,
Who join their heart and join their hand,
   To hurl Sir Jo defiance.

Expulsion! no. As Michael's train
Drove Satan from the heavenly plain,
   To groan in chains of woe:
So Horsley's bays, loud fame shall tell,
Drove the mock president to h–ll,
   To claim a chair below.

Above, however, Banks remained in the president's chair, and Horsley's party – for it was his rather than Hutton's – stood in a hugely frustrating situation. It is clear from the numbers voting at the various meetings that they would have been a majority at any ordinary meeting of the Society. But the new statute about notice of motions meant that they would now always face a house packed with Banks's supporters, and lose any motion they brought in. All was over, though there was a little shouting still to do.

Paul Henry Maty was an unfortunate casualty. He had bravely taken Hutton's part; he had done more than anyone to give Banks the lie direct, contradicting him flatly about his behaviour around certain elections. He had also belatedly admitted that the spark to the whole row was his fault; he had taken away and lost the letter from Geneva that Hutton had been accused of neglecting. His position as secretary was now impossible. There was a bizarre incident on 25 March when he presented to the Society a pamphlet he had written, full of loud criticism of the president, and demanded that Banks move thanks for it. When Banks declined, Maty refused to go on with the meeting. Pressed to do so, he resigned as secretary and walked out.

In another – and mercifully final – doomed sortie, Charles Hutton then offered himself as a candidate for the secretaryship, calling in the promise made to him six years before that the next vacancy would be his. It is hard to see what he can have hoped to achieve. At the vote, Hutton attracted much the same number of votes as he had in December – thirty-nine – and Banks once again got his way; 139 of his supporters voted in his friend Charles Blagden as the new secretary.

Action within the forum of the Society no longer seemed to serve much purpose, and both sides turned to a wider audience. The anti-Banks party collected their letters to the newspapers in pamphlet form in April 1784, and issued another pamphlet based on this material around the same time. As well as Maty's *Authentic Narrative* they produced three more pamphlets over the summer. They provided material to the popular poet Peter Pindar, who quite spectacularly attacked Banks in 1788 with *Peter's Prophecy; or, The President and the Poet*. Hutton himself wrote an anti-Banks pamphlet, but was wise enough to stop short of publication, some said after it had even been printed. The Banks party replied with three pamphlets towards the end of the year. These were strident, partisan: their purpose was to publicise the row and they succeeded admirably. Together with newspaper reports they were sent literally all over the world by Fellows of the Royal Society and their friends. Reports on what was now being called the 'Dissensions' appeared in French and German, and Banks received commiserations from as far afield as India. He was still receiving correspondence from foreign friends about the matter two years later. From his point of view, the visibility of all this material may have been just as damaging as the original row.

He tried hard to play the matter down, telling anyone who would listen that it was a 'trifling affair', a mere spat. And under his guidance, Council decided to award the Copley Medal for 1784 to Edward Waring, Lucasian Professor of Mathematics at Cambridge, for a paper on the strikingly abstruse subject of 'the Summation of Series, Whose General Term is a Determinate Function of Z the Distance from the First Term of the Series'. Thus, Banks felt, the Society demonstrated its esteem for mathematics and mathematicians and refuted the charges brought by Horsley and others that it had become a 'train of feeble amateurs' who viewed the higher reaches of science with uncomprehending hostility. He said much of this in the speech at the end of November when the medal was conferred. Waring must have wondered whether his merits and those of his paper had played any part at all in the award, and in fact he stayed away, sending a friend to receive the medal on his behalf. As an attempt to have the last word in the Dissensions, Banks's stratagem didn't quite work, since his speech was followed by cries of 'Rigmarol!' from the floor, and a proposal to publish it was voted down. It was just a year since Hutton's original dismissal.

∞

Hutton had been caught up in quite a drama. The row had been publicised worldwide, it had brought to a halt one of the most prestigious learned societies in the world and had come close to unseating its president. Much of what took place had, perhaps, been about the private resentments of Paul Henry Maty and the personal ambitions of Samuel Horsley; many had opposed Banks for reasons unrelated to Charles Hutton, and some had supported Hutton for reasons that had little enough to do with Joseph Banks. But when tempers cooled and things began to go back to normal at the Royal Society, one thing was quite clear: Hutton's day in the sun there was irretrievably over.

7

Reconstruction

Walk out of the Academy, leaving behind its bricks and the high windows reminiscent of a schoolroom. Turn right, away from the river. Not to the house in the Warren, but away out of the military riverside site altogether. Leave its noise and its smells behind, and follow the road up the hill. A brisk ten-minute walk, no great climb, towards Shooter's Hill. You quickly gain a view back over the river; there's a faint echo of the climb up away from the river in Newcastle to reach the pit villages: but this walk leads south, into the sun, not north.

Woolwich Common is to your right. Thorns, briars, furze bushes. A few scattered cottages. The local people use the common for grazing, turf and wood. The Academy uses it to test ordnance; cows fall into the holes they leave.

Further up, onto Shooter's Hill itself: actually a section of Watling Street, the ancient London–Dover road. The freshly built Severndroog Castle – triangular tower, hexagonal turrets – and a few little villas for military officers. Although called Ditchwater Lane, it's a deceptively desirable location, in a wooded backdrop to the common. The Warren is hidden by the curve of the hill. And you're home.

∞

Charles Hutton in middle age.

It would be difficult to overstate what a disaster the settled, open dislike of the president of the Royal Society represented, both for Charles Hutton and for those of his friends who had stood by him. Sir Joseph Banks was no pantomime villain. He was an intellectually serious, well-read, energetic man; he controlled what some called a 'learned empire' that took in not just the Royal Society but the Society of Dilettanti, the Society of Antiquaries, the British Museum. He was close to the new political establishment, and to the King; in time he would become George's minister for science in all but name. Many liked and admired him. Banks was most evidently on the scene to stay.

Hutton was faced with the need to do some rebuilding. Neither he nor anyone else was ever actually expelled from the Royal Society – nor did anyone resign – but he could hardly make himself very visible at its meetings after what had happened, and he stopped publishing in the *Philosophical Transactions*, as did a few of his friends.

A few of his friends. Horsley had claimed in January 1784 that there was a substantial body of Fellows ready to secede and form a fresh society. A core group did, indeed, form a club of their own and met on alternate Friday afternoons at the Globe Tavern in Fleet Street, to dine and discuss mathematics and natural philosophy. There was a certain irony to this, because during the previous decade Banks himself had been the leader (the term employed at the time was 'perpetual dictator') of another 'rebellious dining club' that split from the Royal Society's own dining club over a dispute about its rules and met separately for several years. But the new Friday Dining Club had a different agenda, and it aspired, at least at first, to be what Horsley had promised in January 1784, a secession of the core scientific members of the Royal Society, leaving behind at Somerset House a mere rump of dilettantes.

That dream unravelled over the months following the close of the row. Banks's opponents were just not sufficiently united. Some opposed him because they disliked him; some because they disliked his interference in elections; some because they disliked his treatment of Hutton. Some supported Hutton because they liked him, or felt loyalty to him as mathematicians. It hardly made for an enduring rival society, as events would prove.

Indeed, one of those closest to Hutton and most vocal in his support never parted with the Royal Society at all. Nevil Maskelyne, as Astronomer Royal, ran an institution – the Royal Observatory at Greenwich – that was subject to regular visitation and financial control by the Council of the Royal Society, and he therefore had little choice but to remain on speaking terms with its president.

He continued to publish his own papers in the *Transactions* and to pass on to the Society those of others. In the long run his relationship with Banks would deteriorate, but in the short term, during the 1780s, he managed to patch things up quite adequately.

Others did similarly little to keep opposition to Banks meaningfully alive. Paul Henry Maty died within a few years, apparently in penury, having forfeited by his resignation as secretary of the Royal Society his main source of income. Horsley achieved preferment within the Church and became bishop successively of St David's, Rochester and St Asaph, although he continued to publish on mathematical subjects. Francis Maseres, a long-standing friend of Hutton, spent much of the 1790s on a project remarkable for its size, if for nothing else. Inspired by the historical introduction to Hutton's 1785 logarithm tables, he undertook to republish in full all the writings about the construction and use of logarithms mentioned there. The six volumes of his *Scriptores Logarithmici* perhaps predictably burgeoned, filled with material related only tangentially if at all to logarithms – solutions of classes of equation, proofs of theorems in algebra, flattering reprints of some of Hutton's own work – and the series had, even on publication, something of the character of a white elephant.

Hutton's own relationship with Banks, like Maskelyne's, underwent something of a thaw during the decade after the row. Banks visited Hutton at Woolwich during what was probably the summer of 1784, and though the purpose was ostensibly to do with the fledgling trigonometrical survey of Great Britain (later the Ordnance Survey), the wish to mend fences cannot have been far from either man's mind. What was said on that occasion we shall never know, and the same is true of another occasion a few years later when Banks was invited to attend the examination of cadets at the Royal Military Academy. Banks and Hutton even managed to correspond briefly in 1797 – there was a query about the manufacture of gunpowder.

That Hutton himself had other and broader loyalties, ties to more networks than just the mathematical Fellows of the Royal Society, was emphasised in his election to fellowship of no fewer than three foreign scientific societies in the later 1780s: the Royal Society of Edinburgh and the Haarlem Philosophical Society in 1786, the American Philosophical Society in 1788. A colleague from the Paris observatory, Edme Jeaurat, visited around this time, staying with Hutton at Woolwich, talking mathematics and – to judge by a subsequent letter – mildly flirting with his teenage daughters Isabella and Camilla. Hutton had also, back in 1782, joined the philosophical club based at the Chapter Coffee House on Paternoster Row.

A rather sad case on Banks's side of the row was that of Reuben Burrow. Maskelyne's former assistant at the Royal Observatory had received what he felt was inadequate credit for his surveying work on the chilly Scottish hill of Schiehallion, and he never forgot or forgave. Even by 1775 he had come to think of Hutton, Maskelyne and others as the enemy, resenting their prominence and the sums of money their work was bringing them from the Board of Ordnance, the Board of Longitude and the Stationers' Company. Burrow later did a deal with London publisher Thomas Carnan, who had recently won a case at law against the Stationers' Company's monopoly on printing almanacs; Burrows would write and Carnan would publish a rival to *The Ladies' Diary*.

The *Diary* itself had been the site of some mathematicians' spats in its time, but this was perhaps the most absurd of all. From 1780 to 1788 there appeared annually in London two publications with the title *The Ladies' Diary*, each consisting substantially of mathematical puzzles and their solutions. One was edited anonymously by Charles Hutton and published by the Stationers' Company; the other was edited by Reuben Burrow – who was named on the title page – and published by Thomas Carnan. Burrow distinguished the content of his *Diary* from that of Hutton's mainly by including some longer discursive articles on mathematical topics; he never received the flood of contributed material Hutton did, and at times he seems to have been relying on just a handful of cronies, relatives and himself to fill up the pages. But he kept it going, and he inevitably used its pages to keep his quarrel alive, attacking Hutton, Maseres, Horsley and their works. He accused Hutton of stealing his correspondence, and in private judged his *Diary* 'trifling' and claimed he 'does not know how to make an Almanack'. In his own copies of their works he scribbled abuse, some of it obscene. The verdict of Augustus De Morgan, two generations later, was that Burrow was 'an able mathematician but a most vulgar and scurrilous dog'.

Eventually the rival *Ladies' Diary* folded, and Burrow, after filling his personal journal with increasingly violent abuse of London-based mathematicians, took a job in India, teaching mathematics in the Corps of Engineers and later working as a surveyor to the East India Company. He died at Buxar in 1792. His story illustrated to anyone who was interested the disunity of British mathematicians; they were by no means closing ranks and pulling together, even in the face of the threat represented by Joseph Banks.

Burrow's story illustrated, indeed, a trope that was becoming increasingly popular in the sentimental and romantic literature of the time: that a too exclusive attention to mathematics made one argumentative, antisocial, even mad. Radical poet and artist William Blake, in an early satire from the time of the row at the Royal Society, included characters loosely based on its leading players: 'Come said the Epicurean lets have some rum & water & hang the mathematics'. Most closely resembling Hutton was 'Obtuse Angle', a gentle, pedantic man with pockets full of 'a vast number of papers'. He reckoned that 'a man must be a fool ifaith not to understand the Mathematics'. At one point he sang a 'mathematical song' about someone called Sutton. He had a friend (Maskelyne?) called Steelyard the lawgiver.

This was fun rather than savage criticism, but the characters were the ancestors of those in Blake's later personal mythology, in which the imposition of mathematical laws on the world was the act of a malevolent demiurge and his human aides: Urizen wielding his golden compasses. Ultimately this was a view of mathematics and its relationship to the world that Hutton would find himself pushing against for much of the second half of his career, as he pressed his own belief that mathematics was both a useful and a humane part of intellectual culture, a necessary foundation for the sciences and a vital element of any education worthy of the name.

∞

As for practical action against Banks, Hutton and his friends toyed with reopening the debate by demanding sight of the will that created the foreign secretaryship in the first place. The threat alarmed some of those close to Banks, but in the end the plan was abandoned. There was some rumour of an attempt to unseat Banks at an annual election. But again nothing, in the event, was done.

A core ambition was the production of a journal that would announce to the world the seriousness of the new group, thumb its nose at the Royal Society, and of course provide an alternative vehicle of publication for those Fellows who were now unwilling to publish in the *Philosophical Transactions*, involving as it did the scrutiny of a committee led by Banks.

Banks's friend Charles Blagden – now secretary of the Royal Society – was keeping an eye on the Friday Club and its plans. He quizzed Maskelyne, who provided what seems to have been equivocal or contradictory information. In October 1785 Blagden reported that Hutton was busying himself about London soliciting and collecting material for the 'secession transactions', of which he – who else? – would be the editor. There was even a report in one of the newspapers: the 'dissatisfied members' who had seceded from the

Royal Society intended 'to publish their philosophical writings in half-yearly volumes, somewhat in the manner of the present philosophical transactions; by which means the public will still be favoured with the labours of those truly philosophical and scientific geniuses'. Hutton kept a copy of the clipping and probably endorsed what it had said, if indeed it was not he who had briefed the press himself.

It all sounded awfully promising, but it came to nothing. Not one issue of the promised secession transactions ever appeared in print, and there is today no evidence even of what papers or promises of papers Hutton succeeded in collecting for it: if indeed he collected any. The mathematical club meeting on Fridays, meanwhile, carried on until at least 1802, with Hutton, Horsley, Maskelyne, and others including astronomer William Herschel and mathematicians William Frend and Francis Maseres meeting fortnightly during the London season for dinner and conversation. But we hear no more of its ambitions to rival the Royal Society or to produce a journal of its own, and it seems quite clear that it morphed quietly into an unpretentious dining club involving a fairly small circle of friends.

The group that had threatened in all seriousness to supplant the Royal Society and its journal had fizzled out quickly and unimpressively. One factor was that small incremental advances in mathematics and its applications already had perfectly good forums in the philomath journals, including *The Ladies' Diary*, edited by Hutton, and *The Gentleman's Diary*, edited by a man whom Hutton had recommended for the post. The launch of another publication carrying broadly similar material and edited once again by Hutton could hardly have impressed the world and may have failed to impress the London publishers. *The Ladies' Diary*, indeed, was burgeoning at this time, and it's perhaps no coincidence that during the same few years that the 'secession transactions' missed fire Hutton launched – at his own expense – a *Diary Companion*, 'being a supplement to the Ladies' Diary'.

The *Companion* was like a *Ladies' Diary* without the diary – without the almanac that took up the first twenty-four pages. For the rest, the contents were of almost exactly the same kind: riddles, rebuses and 'charades' in verse, mathematical problems and their solutions in prose. The *Companion* carried both supplementary solutions to problems from the main *Diary*, and a separate series of problems and solutions of its own. Perhaps the only notable difference was that, in the somewhat more intimate space of the *Companion*, Hutton did not keep up the pretence of editorial anonymity that still reigned in the *Diary* proper – and he and his contributors were perhaps even less restrained about praising and promoting Hutton's books at every opportunity. Even the elusive Margaret Ord made a couple of appearances as author of verses in the *Companion*.

The *Companion* ran until 1806. It sold well; Hutton was left with just a handful of copies of each issue on his hands, and it seems to have satisfied readers. A few wrote in with their suggestions for its management, which on the whole seem to have amounted to more mathematics and more solutions to each problem. From an initial thirty-two pages it grew to forty-eight.

∞

If Hutton's own impulse to edit a new periodical was absorbed by the *Diary Companion*, at least to some degree, there was still the question of where actual scientific papers from his pen could go. By 1785 he had amassed nine, and the backlog had attained such a size that he took the perhaps obvious course of simply issuing the whole lot as a volume. *Tracts, Mathematical and Philosophical* appeared in 1786, at a time when he still had some thought of making a regular series of 'secession transactions': the title page was that of a standalone single volume, but within the print shop the book was referred to as 'vol. 1' of Hutton's *Tracts*. Indeed, as late as 1795 Hutton still occasionally referred to it as 'my 1st vol. of Tracts', although no second volume would appear.

If a one-off volume was a poor substitute for the ongoing 'transactions' that had been half-promised, it at least contained both mathematics and science of a character that could easily have appeared in the *Philosophical Transactions*. Readers understood that this was matter that would, indeed, have gone to that venue had it not been for the Dissensions.

And the *Tracts* dealt with very much the same range of subjects as Hutton's papers in the *Philosophical Transactions* over the previous decade, giving the clearest indication of the course in which his thoughts were running at this time. There was a new proof of Newton's binomial theorem, the algebraic result that tells you what you get when you raise $x + y$ to a power (any power, even a fraction or a negative number). There were geometrical pieces, including one that concerned obscure new properties of the sphere motivated by questions about perspective: what fraction of the inside of a hemispherical bowl can your eye see from a given point? And a lovely geometrical conjuring trick: divide a circle into several parts that all have the same area and all the same circumference. Finally Hutton gave a new, longer version of his material about ballistics, which he had continued to work on throughout the early 1780s.

It was a fine new mathematical miscellany, and proved to those who cared that Charles Hutton was not dead, crushed or consigned to oblivion. It provided the opportunity for those loyal to Hutton and the kind of British science he stood for to make personal statements of that fact in print; there were kind words in a couple of the book-review journals about the quality of the achievement the *Tracts* represented: full of 'ingenious and useful discoveries'; written with 'perspicuity and elegance'.

As always seemed to be the case with Charles Hutton, there was more. No sooner had the *Tracts* appeared than he was firing off a couple of papers to the Royal Society of Edinburgh. One dealt with experiments on the expansion of freezing water reported to him from Quebec by his old friend Edward Williams, now a major in the Royal Artillery (where there was plenty of freezing water, and you could make an iron shell burst by filling it with water and leaving it outside all night). Another reported Hutton's own ongoing work on the resistance of the air, a topic that would increasingly exercise him over the next decade. More privately, Hutton was exchanging manuscripts with friends on topics including the motion of comets and the vibration of strings. He occasionally carried out an observation at the Royal Observatory, under the presumably watchful eye of his loyal friend the Astronomer Royal. He was reading furiously as ever in British and foreign mathematical works. At the same time he was developing his interest in the history of mathematics: several of the *Tracts* touched at more than token length on the history of the problems they attacked, and this, too, would become one of Hutton's obsessions into the 1790s.

That historical interest came together with the continuing need to rebuild his career and his public persona in another curious habit Hutton acquired around this time: that of discovering that he was related to other minor (or sometimes major) celebrities. It was something he talked about rather than wrote down, but by the time of his death it could be stated in print as an uncontroversial fact that he was the cousin of the nonconformist leader James Hutton, although the latter's biographers were unaware of the fact and it has so far defied confirmation from the genealogical records. At some stage he and the popular novelist Catherine Hutton also decided that they were 'cousins', though in this case there was an air of acknowledged fiction about the matter. One biographer was confused enough to invent a specific relationship (daughter of mother's sister), but it was eventually acknowledged in print that the two families were not really connected.

Most spectacularly, Hutton started putting it about that he was related to the *ultimus heroum* of British science, Sir Isaac Newton. Newton never married and had as far as anyone knows no children, so the closest relationship available was to be descended from Newton's mother. By 1805 the story had acquired such currency that Lord Stanhope left Hutton in his will the famous Vanderbank portrait of Newton, as being the most fit person to have it. A descendant eventually passed it on to the Royal Society, where it more properly belonged, but the tale continued to circulate in the form that 'cousin' James Hutton's grandmother was the sister of Newton's mother. (There is, to be fair, evidence that James Hutton really was a descendant of Hannah Ayscough.) Fully elaborated, the tale had the Hutton family hailing originally from Westmorland, with one branch then moving to Lincolnshire (where the Newtons lived) and another to Northumberland. Entirely beyond proof or disproof today, the story is a testimony more than anything to Hutton's assiduity in persuading people that he belonged in the world of metropolitan science to which he still – and since the Dissensions more urgently than ever – aspired.

A more solid piece of personal reconstruction came in 1785. In May that year Hutton's wife Isabella died in Jesmond near Newcastle. She was buried in the Dissenters' burial ground in Percy Street: very close to Hutton's birthplace, as it happened. The mails being what they were, it's probable Hutton was unaware of her death until the funeral was over; certainly he did not attend or visit the north at this time.

He lost no time in regularising his relationship with Margaret Ord. She became Margaret Hutton just two months later, in what was evidently quite a private ceremony at the Fleet Street church of St Dunstan in the West. It was the parish church for the set of rooms Hutton rented at Clement's Inn, and both parties gave Clement's Inn as their place of residence. There must have been a degree of connivance on the part of the celebrant, Joseph Williamson, who presumably knew perfectly well that neither party really lived in his parish.

Charles and Margaret's daughter Charlotte was sent to school at a convent in France around this time, together with her (half-) sister Eleanor. Hutton was keen on French culture ('it is but justice', he wrote in the *Tables*, 'to remark the extraordinary spirit and elegance with which the learned men and the artisans of the French nation undertake and execute works of merit'). And evidently he was keen for his children to acquire a high degree of social polish: another aspect of his rebuilding and self-reinvention during this period. But just what prompted what was surely a rather odd decision for an English Protestant is one of Charles Hutton's mysteries.

∞

The 1780s were a period of reconstruction not just for Hutton personally, but also for the country and the institution for which he worked. Victorious in most of the American battles, Britain had nevertheless lost the war, and with it the thirteen east coast colonies which now formed the United States of America. George III received John Adams as Minister Plenipotentiary in June 1785:

> I was the last to consent to the separation; but the separation having been made and having become inevitable, I have always said, as I say now, that I would be the first to meet the friendship of the United States as an independent power.

Britain was demoralised, humiliated, and the war had been costly in monetary terms. Servicing public debt was costing around two-thirds of public revenue, and cutbacks were inevitable.

After the Treaty of Versailles in 1783 the British Army was sharply reduced in size. The Royal Artillery, too, was cut by more than half, from over five thousand men to just two thousand. But twelve companies were still stationed at Woolwich, and the cadet company was left untouched. After years of frantic need for officers there were, suddenly, no commissions in sight: for four years no cadets were made lieutenants. Public exams restarted and there was a sense of normality returning, at least as far as the teaching was concerned.

The cadets were now arranged in three 'academies' within the Academy – a 'second' was added to the old upper and lower academies – and each of those into four classes. Judging who had achieved what and who deserved to be moved up or down took a good deal of effort: written class lists and certificates from the staff; examinations to progress from one academy to another. Such a degree of fuss and documentation was unknown in normal schools. Hutton's role and judgement remained crucial, and mathematics – algebra and geometry – remained prominent in the teaching, the tests and the examinations.

It was in some ways much the same as before the war: the same routine of calling out boys singly or in small groups to talk them through their work while the rest got on with exercises. The same round of definitions, rules and worked examples, pulled from book after book; selecting and arranging and editing the material into something uniform enough to make sense. Hutton wasn't one to stick rigidly to a printed source, even if it was such a revered text as Euclid's. When the students complained about parts of the theory of ratios in the *Elements*, he dropped or modified it.

He got on with the cadets, by all accounts. Howard Douglas, who joined in 1790 aged thirteen, was a quick favourite of Hutton's: head of his mathematics class and so proficient that Hutton used him as a sort of assistant, telling boys in difficulty with the material to 'go to Douglas'. He always remembered Hutton with gratitude and affection. Another commentator noted Hutton's 'goodness and simplicity' towards both his colleagues and the cadets in his charge; one of the junior masters under Hutton later wrote that

As a preceptor, Dr. Hutton was characterised by mildness, kindness, promptness in discovering the difficulties which his pupils experienced, patience in labouring to remove those difficulties, unwearied perseverance, and a never-failing love of the act of communicating knowledge by oral instruction. His patience, indeed, was perfectly invincible. No dullness of apprehension, no forgetfulness in the pupil, ever induced him to yield to irascible emotions, to forfeit his astonishing power of self-control.

. . . during the last twenty-five years, I have had the most favourable opportunities of acquainting myself with the best modes of giving instruction in the University of Cambridge, and in other institutions, both public and private; and during much of that time, I have been extensively engaged in the same profession; but I do not hesitate to say that I have neither seen, nor have the least conception of, any oral instruction, the excellencies of which bear any comparison with those of Dr. Hutton.

∞

The immediate post-war period saw one final piece of rebuilding from Hutton. He had always complained of the unhealthiness of the Woolwich site. Miasmic in the summer, it could be bitterly cold in the winter, the wind off the river blowing chill down the long open spaces between the buildings. He claimed in print that the air at Woolwich had shortened the lives of two of his predecessors.

In May 1787 Hutton's complaints about his health came to a head, and he applied for leave to move further away from the Academy than the house in the arsenal he had inhabited so far. He was convinced that the location was responsible for his persistent lung complaints, though early inhalation of quantities of coal dust probably played a role as well. He represented that his health was in danger from this and perhaps from overwork.

He was granted house-rent and he moved up to Shooter's Hill. The air and the exercise of climbing the hill (twenty minutes or so, done briskly) did him good. There were a few other officers' houses there: possibly a certain amount of society for his family. If the Royal Artillery sometimes pounded away on the common below, it was equally often Hutton's own experiments doing the pounding, so he could scarcely complain.

Neither could it escape him that the common itself was ripe for development. From 1774 building plots there had been sold off, a plot having been enclosed a year before for the new Royal Artillery barracks. Meanwhile a Barrack Tavern was added to Shooter's Hill, and one of the large houses became a small factory. When ten acres of freehold on the common came up for sale, Hutton had little hesitation in bidding, and after delays which in the end stretched on for several years, he succeeded in buying.

He cleared the existing low cottages and in 1790 began to build. He had designed and built a house in Newcastle already, and he was an authority in print on at least the surveying, quantity surveying and accounting side of house building. It's probable that the designs were his own, and certain that he oversaw the work closely. Even the bricks and slates were manufactured locally, from the clay of the common. Hutton interested himself in the process, and was said to have hit on certain improvements and even to be setting up in business as a brick manufacturer, though that came to nothing.

It was said, too, that there was no water on the site, but Hutton obtained a set of iron boring rods and by drilling down fifty feet found a spring and dug wells. All this cost money, and Maskelyne lent him a hundred pounds in August 1793.

There was a certain, perhaps inevitable eccentricity to what resulted. Echoing the geometrical folly that was Severndroog Castle, Hutton built for himself and his family what came to be known as the Cube House. Its foundation was a forty-foot square, and it was forty feet high. It was not wholly austere, however: Ionic pilasters and bow windows both north and south made it a handsome villa, and he stuccoed it.

He built more, in more sober style: pairs of semi-detached houses in two rows, some of four storeys and some of five. By 1805 almost twenty of the houses that Hutton built stood in a sort of suburban park estate, a private village. Hutton and his family lived on the site for a number of years. And of course, after the initial outlay of capital, the rents brought him a comfortable independence. By the time he was paid back, Maskelyne had made a profit of 30 per cent on his loan; Hutton certainly made more, and it got about that the canny mathematician had made his fortune in the building business.

8

A Military Man

Woolwich Common, 13 September 1787. A fine dry day.

Sponge your gun. Load with cartridge; with ball; ram. Run up the gun; traverse it to aim. Prime and fire.

Bang.

When the smoke clears the assembled class of cadets groan, as perhaps do Major Blomefield and Professor Hutton. Not for the first time, the firing crew has missed its mark, failing to hit the ballistic pendulum from seventy-eight feet away. Eighteen ounces of iron, travelling slightly faster than the speed of sound, have buried themselves harmlessly in the earthwork beyond.

Reset the pendulum and try again. Sponge; load; ram; aim; fire.

Hutton taught at the Royal Military Academy for twenty more years after the end of the American war and the shipwreck of his career at the Royal Society, and he achieved a great deal there. It was being said, and it would continue to be said, that he had raised the standard of achievement at the Academy quite strikingly. 'Our British university for the military science', one magazine commentator called it, indicating the degree to which it now exceeded what was expected from a mere school. Admiring quotes are cheap, but a clearer indication of the high reputation of the Woolwich teaching was its use as a model by other military schools. Private academies adopted its textbooks and in some cases imitated its weekly schedule of classes. The foundation of the Royal Military College, initially at Great Marlow and High Wycombe in 1801 – it later moved to Sandhurst – was avowedly on the model of the Academy at Woolwich, with three departments corresponding to Woolwich's three 'academies', and a curriculum based from the first on the study of mathematics. It was in a sense a satellite of the Royal Military Academy, and the same was true of the East India Company's training college, founded in 1809 at Addiscombe. There the Woolwich Academy was a model in respect of uniform and badge as well as in the style of teaching; there had long been links between the training of East India and Royal Artillery cadets, the former actually being accommodated at Woolwich for a period.

Thus, Hutton's mathematisation of the syllabus was no local phenomenon but the beginning of a change in British military training that would have profound effects on the mental attainments and habits of mind possessed by British officers, not just in the Royal Artillery and the Engineers but across the entire Army. In time, his ideas would spread also to North American military training and to British India.

To teach well was a fine thing, but to export one's ideas on this scale involved more, and Hutton was once again engaged in translating his knowledge into textbooks during the 1780s. As with his round of education-based writing in the 1760s and 70s, this would support his teaching, removing the need to bring several different printed sources into the classroom; it would also, of course, consolidate Hutton's own reputation. And the captive market represented by the Royal Military Academy and the other military schools meant that such ventures could be confident of making a profit, too.

First came a sort of junior version of his *Mensuration*, the 1786

Compendious Measurer. 'Brief, yet comprehensive', it was intended for school and practical use; as a result it adopted a more familiar manner and was much smaller and cheaper. There were no proofs; this was simply a compendium of rules and examples suitable for study in the classroom. Definitions were cut down to a minimum; each rule was given in a form suitable for copying into the student's own book; each had a worked example and a diagram, and each had questions with answers.

The *Measurer* was, for much of its length, a curiously bland book. The arithmetical part had no real-world examples except where absolutely necessary: there was none of the colour and little of the idiosyncrasy of Hutton's earlier textbooks such as the *Guide*, and much less sense that Hutton's own personality was on display. Very many cases of measuring solid shapes were worked through without any indication of when you might meet them in reality or how you would know if you did. There were no prisms that resembled hatboxes. But the book did swerve towards practicality at the end, with sections on a wide range of real-world applications: measuring timber, brickwork, marble slabs; carpenters' and joiners' work, slaters' and tilers'. Plasterers, painters, glaziers, pavers, plumbers: you can almost hear the roll-call of people who had worked or were working on Hutton's own building projects up on Woolwich Common, and it's tempting to wonder whether any sections of the *Measurer* used, unacknowledged, advice from Hutton's workmen at home. In a telling slip, he referred in one problem to 'my plumber'. There was a military flavour too, with questions about the weight of a pint of gunpowder or the best way to pile up spherical cannon balls. And, again, a keen sense of the physical space in which Hutton worked: 'How long, after firing the tower guns, may the report be heard at Shooters-Hill, supposing the distance to be 8 miles in a straight line?' (the answer is $37^1/_3$ seconds).

There was a section in the *Measurer* on conic sections, those shapes you get when you take a slice through a cone (I always imagine a conical piece of cheese cut on a slant by a knife): ellipse, hyperbola, parabola. After a brief interlude in which he published a *Key* (a book of worked answers) to his *Guide*, Hutton worked up this material into a book-length treatment of the *Elements of Conic Sections*. Which version came first – the long one in the *Conics* or the short one in the *Measurer* – is hard to say; they had a lot in common in the way they were organised, and quite a lot of straightforward repetition of questions; quite obviously they had their common source in Hutton's teaching at Woolwich.

The *Conics* was a hard book – perhaps Hutton's hardest – and it was written in a definition–theorem–proof style that he elsewhere usually avoided. Proofs ended with 'QED'. He set out steps of argument on separate lines for ease of view, but by doing so he added to the sense of almost oppressive formality in this presentation of difficult material. He also, as far as possible, reused the same wording in the three major sections of the book – devoted to ellipses, hyperbolas and parabolas – heightening the impression of an almost hieratic formality and stateliness of presentation. For me it conveys less the delight in his own dexterity of his other books than mere slog. It doesn't help that the odd mistake can be found, resulting from the reuse of text without all the necessary changes. It's hard to imagine what the cadets made of it. Admittedly conic sections, notably the parabola, had some importance to the theory of artillery, but the students can surely have felt little real enthusiasm for it. The austerity was leavened somewhat at the end of the book with – as in the *Measurer* – a random selection of practical exercises, many of them taken from the *Mensuration* or the *Measurer* and some of them quite fun. Repeat the working that led Archimedes to cry *eureka*. If you're above London in a hot air balloon and can see Oxford, how high are you? How fast must a 32-pound cannon ball move to do the same damage as Vespasian's celebrated battering ram?

For all that, the *Conics* was notably well received. It circulated as a manuscript for a while, and was taught from in that form; eventually it received the commendation of Charles, Duke of Richmond, now the Master-General of the Ordnance. Ordered to be printed, it appeared in 1787. Mathematicians admired it: 'I am much pleased with your Conicks', wrote John Playfair, 'and expect much advantage both from them and from the collection of Problems at the end.' And, presumably at the instigation of Richmond, Hutton received a perhaps singular honour for a mathematical author; he was presented at court and allowed to kiss the King's hand. It's a pleasure to imagine the pit boy at the court of St James, and the scientifically minded king, patron of Sir Joseph Banks, meeting, however briefly, the pariah of the Royal Society. The two men were near-contemporaries, Hutton just a year older than his sovereign.

A man who could be presented to George III for writing a two-hundred-page textbook about conic sections was a man who was good at more than just mathematics. The incident showed Hutton at his very best as a skilled manipulator of networks and user of personal connections. At a less frenetic pace than in the 1770s, perhaps, and with less of the heroic incaution of his younger days, he was building up a rich world of people who liked and admired him, people who owed him something from having been taught, helped or advised; and prestige followed, as indeed did the more tangible rewards of money and employment.

As well as teaching and publishing, Hutton had never abandoned his dreams of being taken seriously as an experimenter, a natural philosopher, and he returned to the appropriately military subject of ballistics with new seriousness after the American war was over. He had continued to make experiments since his 1778 Copley Medal, with a new long account appearing in the 1786 *Tracts* while the experiments were still going on. Ultimately he staked his hopes of scientific greatness on the subject, and expended a great deal of effort on making experiments over the years from 1775 to 1791. By the end he had blown up nearly a ton of gunpowder in the pursuit of knowledge.

The problem was an old one. A body fired in a constant gravitational field travels in a parabola if air resistance is ignored, and you can describe the parabola and figure out exactly where it will land if you know the speed and the angle of elevation with which it left the gun. Range tables calculated from this theory had been appearing in print since the sixteenth century, and gunnery teaching that used it still throve in private and public academies across Europe.

Unfortunately, air resistance cannot be ignored. Working gunners knew perfectly well that the parabola theory therefore provided no description at all of where real cannon balls really fell, with the result that these large, expensive, dangerous pieces of equipment were being employed on a trial-and-error, rule-of-thumb basis. If the target was close enough you pointed your gun straight at it (in other words, point-blank range). If not, you pitched the gun up a few degrees and adjusted until you hit it. Range tables, to be any use, had to be worked out empirically for the individual gun. For the purpose of firing mortars or the new exploding shells at high elevations, say into a besieged town, this simply wasn't good enough, and the knowledge that the theories they were teaching and learning were incorrect was demoralising, to say the least, for students and staff in institutions like the Royal Military Academy.

Several things were going wrong, and Hutton hoped to elucidate all of them by a sufficiently rigorous experimental study of cannon and the projectiles they fired. For a start, muzzle speeds simply weren't known. Hutton's work with the ballistic pendulum, extending that of Benjamin Robins, had begun to change that. His experiments enabled him to say with increasing confidence how muzzle speed depended on the weight of the powder charge, the weight of the shot, and the length of the gun.

So far, so good. Another problem for the practical gunner was that the strength of gunpowder was variable. With a given charge, sometimes the ball would blaze out of the muzzle at a thousand feet per second and sometimes it would do rather less, emerging in extreme cases with a mere pop. So Hutton invented – improved might be a fairer way to put it – a machine to test and quantify the strength of gunpowder, and in his own experiments he put in place a system of sifting out the coarser, less effective grains from the powder until what was left had the uniform strength he required. Another problem solved, and Hutton's 'eprouvette' found practical use later on for testing gunpowder at the Royal Arsenal.

A third issue was air resistance, and its complicated dependence on the speed of a projectile – which itself was constantly changing in flight – as well as its shape. As well as firing balls at pendulums, Hutton also measured ranges directly by firing balls down the Thames at elevations of fifteen degrees. He stationed two parties of cadets to note the splashes where the balls fell. (No one seems to have warned the shipping.) The results were of some use, although a mathematical law would continue to be somewhat elusive. But the work was profoundly demoralising in another way, because it made it quite clear that firing balls at long ranges could not be an exact science. In a mile's flight the shot were wandering anything up to a few hundred feet to the left or the right. The skilled gunners of the Royal Artillery were unable even to hit the river consistently. What was going wrong?

Hutton blamed, in part, the quality of 'windage', which he had already shown was wasteful as to the use of powder: the difference between the size of a ball and the size of the bore, which for safety's sake was often set as large as a twentieth of the calibre of the gun: a tenth of an inch or more. With the ball rattling from side to side in the barrel by that much, he reasoned, perhaps it was no surprise it didn't always go where expected. He wondered whether cylindrical shot might fly straighter, as well as advocating a reduced windage to control the problem.

But, once in the air, there might be wind to deflect the ball, and as Robins had already hinted there seemed to be a distinct tendency for spinning balls not to fly straight but twist away to left or right even in calm conditions (the phenomenon would eventually be studied under the name of the Magnus effect; it's real, and it makes spherical cannon balls irretrievably inaccurate at long ranges). All this tended to point to the conclusion that however many experiments Hutton did and however good the mathematical laws he derived, they wouldn't enable gunners to hit anything they couldn't hit already.

Yet still he persisted. Direct measurement of ranges proving less conclusive than he might have hoped, Hutton investigated the resistance of the air more closely, hoping to discover how much speed a given ball lost in travelling a given distance. There were some mathematical models and some good guesses in print, but Hutton hoped he could do better. It was a problem Newton himself had worked on, and it would have been a coup indeed to solve a problem on which the Newtonian theory was known to be incomplete.

So he fired balls at a pendulum from a cannon that was itself mounted on a pendulum. By measuring how much the freely swinging cannon recoiled, he could deduce the ball's muzzle speed; by measuring how much the ball-struck pendulum was displaced he could deduce the ball's speed at impact. The difference between the two told him how much speed the ball had lost due to air resistance.

The method worked, and Hutton pursued a lengthy series of experiments enabling him to say in some detail how the air's resistance depended on the ball's size, weight and speed. Not satisfied with this, he also found some of Robins's old apparatus and used it to do more work on air resistance. The 'whirling machine' placed an object at the end of an arm, and used the force from a descending weight to whirl it around in a big circle. Descending weights behave predictably, and by observing how fast the weight was falling compared with freefall, Hutton could deduce how much the object on the arm was resisted by the air. It was a neat, delicate apparatus, and it helped Hutton fill in gaps in what he knew about the resistance exerted by air.

Other people were also interested in ballistics experiments, and Hutton received detailed reports of similar work to his, done by officers at Landguard Fort and at Gibraltar; he also saw reports in the *Philosophical Transactions* of experiments by Count Rumford, of which he did not think much. He briefly attempted a synthesis of some of the data that had come his way from these various sources, but quickly abandoned it, perhaps reasoning that he did not wish to rely on experiments done outside his own meticulous control. The work at Woolwich went on.

But by the end of summer 1791 it was all too clear that, overall, the results were disappointing. Air resistance obeyed no clear mathematical law. Yes, it depended on the surface area of the moving body, and different shapes (a cone or a hemisphere, say) produced consistently, quantifiably different degrees of resistance. But the key question was how resistance depended on speed, and here Hutton was unable to make things work. He massaged the data to 'regularise' them (equivalent to drawing a straight line through a scatter plot); he laboriously fitted equations to the data once regularised. Because of the physical model of air resistance he had in mind, he thought that above a certain high speed one component of the resistance would reach a maximum and remain constant. But because he wanted a single equation to work with, he instead sought one law that would cover all speeds. He found it in the rule that resistance was proportional to the two-and-one-tenth power of speed; but because that was intractable to work with (in calculus terms, he was unable to integrate it) he abandoned it in favour of a different rule, a sum of multiples of the speed and its square, that fitted the data much less well: one of the saddest moments in the whole work.

Again and again Hutton's frustration was obvious in his writing about the ballistics experiments. Now he had fixed on a mathematical law describing the resistance, he would have liked to be able to use it to derive a description of the shape of a projectile's path. But that proved intractable for purely mathematical reasons. He drew a picture and gave a description in words, but he could find no way to describe it quantitatively, try as he might. And he would have liked to make some theoretical predictions for the range of different balls at different speeds; but, similarly, the equations wouldn't or couldn't be solved.

Much had come out of the ballistics work that was quantitative, robust and useful: laws for muzzle speed in relation to shot size, powder charge and gun length; a law for the ranges of shot fired at 45 degrees of elevation; a law for air resistance in relation to size and speed for three different shapes of projectile. All of these had physical models attached that did at least something to explain or justify them, though they were not the most convincing or the most interesting aspect of the work. There were interesting side issues such as how long a cannon ball would take to attain a given height and how far it would go horizontally in a given time; or how high it would go if you fired it straight up. But putting it all together eluded him; the crown jewels of how air resistance operated on a projectile whose speed was changing, and how as a result the range of shot depended on the speed at which they were fired, escaped his grasp.

Hutton had missed his aim, and his work on ballistics had not quite achieved the key results that would have made his scientific name for ever. It is very hard indeed to say how much his work contributed to military practice. Certainly he had found some definite results about the speed of balls and how far they would go in different circumstances; perhaps the most interesting and useful item was his calculations of the size of charge that would achieve the maximum velocity from a given gun. Certainly he taught all he had found out to the cadets under his charge. He summarised his results in exercises that he included in his text-books, and in tables that could in principle have been of use to gunners. The work was much praised: 'Military tactics have been much benefited by his important labours', one obituarist would say, and some of the textbooks of gunnery of the next generation gave his work and his results a lot of space. There seems at least an indirect connection between Hutton's demonstration that longer barrels added little to muzzle velocity and the enthusiastic intro-duction of short-nosed 'carronades' in this period; and in the fullness of time his opinion about reducing windage would become the accepted one.

Perhaps more important, if less direct, was the impact of Hutton's ethos of experimentation and military improvement on the cadets and staff with whom he was in contact. There passed under his teaching Henry Shrapnel and William Congreve, who would become the inventors of the shrapnel shell and the Congreve rocket. New types of rifle were tested at Woolwich; experimental work on the strength of timber was carried out. Chemical experiments took place in the Royal Laboratory, with outcomes for the quality and consistency of gunpowder and the methods by which it was tested, a subject to which Hutton himself had contributed with his new design for an eprouvette for the purpose. The whole arsenal site was a semi-public place that keen teachers and students visited to see experiments in progress; and indeed there were those who reckoned it an important site for the now-gathering reform of British science: dynamic, innovative and daring, and away from the eye of the Royal Society.

∞

By the time Hutton's work on ballistics was finished war was once again threatening. An unhappy consequence was that it became impossible to publish the sensitive military information his work contained. Summaries appeared in his textbooks, but the masses of data and the full reports, drafted and redrafted, remained for years in manuscript form, circulated within Woolwich and the Board of Ordnance; they were not printed in full until the war was nearly over.

Just ten years had been allowed to Britain and its armed forces before they were once again plunged into destructive war. The French invasion of the Netherlands, and the execution of the French king Louis XVI in early 1793, brought Britain into war against revolutionary France, and for most of the next twenty-two years Britain was at war with her neighbour in some combination of alliance with or against other continental nations and (from 1812) America. On the one hand the fear of invasion (scares in 1796 and again in 1803–5; two actual landings in 1797 and 1798) and a shared Francophobia made it a stirring, patriotic time that – some say – helped forge a nation. On the other, the spectacle of the French Revolution and of the military violence that followed were persistent sources of disorder both political and physical, and of deeply felt division and unease.

By 1793, too, nine years of economy, reorganisation and reform under Pitt the Younger had brought the national debt under control but left the Army in no condition to fight. The armed forces and their administration were obliged to expand massively during the early years of the war, and a whole breed of career administrator sprang up to keep the system operating. A massive war, a massive war effort; one man in four of those of military age bore arms either as a regular or an auxiliary, through enlistment or balloting (the country stopped short of actual conscription). Professional families flooded the officer class; the business and trade classes the lower ranks. And in some campaigns 30 or even 40 per cent of those serving died. Britain lost perhaps 300,000 men in the Napoleonic wars.

The experience of the Royal Artillery in the Napoleonic period is encapsulated in the award in 1833, after it was all over, of the motto 'everywhere' (*ubique*) to the regiment. Suddenly William Congreve was not just an experimenter and director of the repository; he was commander of the artillery in Flanders in 1793–4. The shrapnel shell and the Congreve rocket were no longer exciting toys but devastating new weapons. The boys who had carefully aimed shots into pendulums and over the Thames with Hutton fired ten thousand rounds in a day from their seventy-eight guns at Waterloo.

Like the Army, the artillery regiment was expanded; this was a war that made huge use of artillery, the great opponent Napoleon being in origin an artillery officer of supreme tactical skill. A new horse artillery was created, followed by new battalions of regular artillery every few years, taking a strength of four battalions of artillery at the outbreak of the war up to ten in 1808; from four thousand men to twenty-seven. Much the same was true of the Royal Engineers, who served with extraordinary versatility in work ranging from sieges to river and road communications and river crossings, surveying and reconnaissance. The training provided by the Royal Military Academy to gunners and engineers was suddenly – once again – desperately important.

And once again it was simply impossible to provide that training to enough cadets fast enough as the demand for officers grew. The number of cadets increased, reaching officially two hundred at the peak: most of them were at Woolwich, though from 1803 some were based at Great Marlow. There were complaints – once again – about overcrowding, about poor sanitation. Scarlet fever broke out. Hutton in 1795 wrote that the lecture rooms and barracks at Woolwich were 'so small as to be insufficient for the purposes of the institution'; he also reckoned that the salaries of the staff were inadequate to their labours, a situation to which inflation after the American war had contributed, as had a ban on supplementing one's income by taking private pupils. In 1797 the staff wrote jointly to the Board of Ordnance to ask rather desperately for a pay rise, citing wartime price inflation and their wish 'to support ourselves and families with . . . credit and decency'; and quite surprisingly they got one.

They were indeed teaching at a furious rate. A deal was signed that brought the cadets of the East India Company into the Woolwich system, swelling numbers in the regular classes. Furthermore, the Irish corps of artillery merged with the British in 1801 and came to Woolwich to acquire theoretical knowledge, attending the professors at their houses on a total of four days a week for three hours each day. 'One guinea for each lesson to three pupils, or any less number, and seven shillings per lesson for every pupil more than three who may attend for instruction.' Hutton also undertook to lecture on natural philosophy from May 1799, on top of his mathematics teaching.

In 1803 the summer vacation was simply cancelled 'in consequence of a most serious want of officers for the Royal Regiment of Artillery'; by the end of that year seventy-seven of the hundred cadets in the upper academy had received commissions. Formal exams were once again totally abandoned, and a sense arose anew among the boys that you were likely to be commissioned no matter what you did. 'Disgraceful irregularities' are heard of in the records, 'unmilitary and riotous behaviour'. Cadets had to be forbidden to dine at the officers' mess, following an incident in which 'Mr. C— . . . was so drunk that he knew not what he did'. Other misdeeds ranged from insubordination and cheating to shooting a local woman with a fowling piece, and in December 1795 the Master-General of the Board of Ordnance remarked, appalled, that 'this Institution being intended for Gentlemen only, the Regulations have not provided punishments for an offence, which it was supposed no gentleman could commit'.

The curriculum was abridged in practice if not in theory, and eventually, in 1803, two new mathematical assistants were added to the staff to cope with the increased numbers, 'it appearing to the Master-General that the Professors and Masters of the Royal Military Academy bear no proportion to the pupils under their instruction'. In 1806 three more assistants were added, and by then the system involved no fewer than six numbered academies, each subdivided into graded classes. It was becoming clear that the management of students' progression from one class to another was not working properly: too complex, too burdensome. Hutton's role in charge of at least the mathematical side of curriculum and organisation was the pedagogic equivalent of high command, and it stretched his skills. In the long run it would be said that he let things ossify under his rule, with the students merely cramming rather than being properly taught. But still the demand for officers pressed, and there was little he or anyone else could do.

For all the chaos, it seems quite clear – as it probably was to most of those involved – that the teaching at the Royal Military Academy was of real utility to the war effort and the country. Until the Royal Military College was founded at Great Marlow there was no other major source of officers with technical training, and the training was far in advance of anything a regular Army officer could be expected to have. It was indeed a comprehensive scientific education, producing officers who knew their trade when commissioned and could function at once, without needing more experienced officers to help and guide them. Graduates of the Royal Military Academy successfully did work ranging across surveying and reconnaissance, liaison and engineering support, translation and the full range of ADC and staff-officer duties; made themselves invaluable in theatres of war from India to the Caribbean.

∞

The chaos of war was accompanied by changes to Hutton's own household. By autumn 1790 his daughters Charlotte and Eleanor had come home from France (perhaps in some haste; we don't know exactly when), rejoining the older girls Isabella and Camilla. Charlotte was a particularly welcome addition to the Cube House; she had become a young woman of charming temper and remarkable intellectual gifts. Early to rise, she helped to manage the house and the servants and was a ready conversationalist at Hutton's often-visited table. Witty, cheerful, charming: but more, she alone of Hutton's children showed signs of inheriting his talent for mathematics.

Like her mother before her, Charlotte assisted Hutton with calculations. She made a new expanded table of square roots (to twelve places of decimals) for inclusion in one of the many re-editions of his works. She made drawings and maps for her father, and took over the arranging of his voluminous library. There is little doubt she was Hutton's favourite, and in the midst of war the Cube House was for a while a lively and a happy place.

One visitor had a more transformative effect than most. It had to happen sooner or later, and it was Camilla, it seems, who was the first of Hutton's daughters to be seriously courted. Charles Henry Vignoles was the fourth child of an Army officer, and a junior officer himself. Quite how he became acquainted with the Hutton family we don't know, but by October 1790 he was finding himself tongue-tied in the presence of his beloved's father and resorted to a letter, in which he set out with perhaps excessive frankness his own personal financial situation: he was in debt to his friends for the price of his adjutancy. Without directly asking, he made Hutton's consent to the match conditional upon Hutton's settling the debt himself. With a possibly surprising magnanimity Hutton did consent, and the pair were married a week before Christmas.

Soon after, Vignoles was posted to Ireland, and Hutton was

informed by letter of the birth of his grandson Charles Blacker Vignoles in County Wexford in 1793. But then Vignoles's regiment was ordered to the West Indies. The struggle with France for control of the Caribbean islands had made the region an important theatre of war; it was also a most notoriously unhealthy station, with a deserved reputation for yellow fever. Camilla nevertheless determined to accompany her husband, taking baby Charles with them. Hutton was not happy, and there seems to have been something of a row about the matter, but there was nothing he could do. The couple embarked at Cork in November, bound for a combined operation against the French islands of Martinique, St Louis and Guadeloupe, to be based in Barbados.

Isabella and Camilla Hutton.

9

Utility and Fame

The twenty-seventh of September 1794. The church of St Luke, Charlton: Charles Hutton's parish church.

Bitter wind; falling leaves. The long view north: past the Warren, across the river and over the levels. Tears dim the distance.

A new-made grave; earth cast into it. *We commit her body to the ground; earth to earth, ashes to ashes, dust to dust; in sure and certain hope of the resurrection to eternal life, through our Lord Jesus Christ.*

In his final decade at Woolwich Charles Hutton would know both triumph and, repeatedly, tragedy. By the 1790s, he was famous. Two brief biographies appeared in print. You could refer to 'Dr Hutton' and – as long as the context was mathematics – you would be understood. Newspapers and reviewers were calling him 'distinguished', 'learned', even 'veteran'. *The Ladies' Diary* in 1797 included an acrostic puzzle whose answer was 'Hutton', providing an opportunity for those who knew he was the editor – and a few who didn't – to utter their homage:

. . . a mathematic sage,
　　To him what praise belongs!
Whose works improve and grace the age
　　And shine in Pindar's songs.

. . .

Whose fame doth reach the arched skies,
　　Whose works we all admire.

. . .

　　The work of his own hand,
As long as learning is belov'd,
　　Will unimpaired stand.

There was a sense at least in this community that all his efforts to recover from the disaster of 1784 had borne fruit, and that his status was as high or nearly as high as it had been before. The *Tracts*, the long-running textbooks and the widely used tables had made his as close to a household name as a mathematician could get.

A measure of his fame was a growing tendency to consult him on mathematical matters outside his areas of real expertise. On several occasions from the 1790s onwards engineers, architects and public bodies sought Hutton's opinion about bridge-building projects: a stone bridge at Sunderland; two iron bridges in Wales at Conway and Bangor; repeated proposals for a new London bridge. Hutton was neither engineer nor architect and had never designed or built a bridge in his life, yet he evidently had enough of what would now be called name recognition for his opinions, impressionistic as they sometimes were, to be worth something. He had become the man non-specialists most readily thought of when they thought of mathematics.

Hutton always said he had aimed not at fame for its own sake but at public utility, and as the years passed he protested more and more strongly that his only desire was to be of use to his country. There were those who believed him, but it is difficult not to detect, in some of his activities, at least the supplementary desire to build monuments to Charles Hutton. Nowhere is this more true than in the series of massive publication projects that would appear in the decade after 1795. After the 1787 *Conics*, in fact, Hutton never again published a book that consisted substantially of new material; every later project was based mainly on editing or reworking of one kind or another, although some large chunks of original research were, admittedly, included along the way.

Even before the *Conics* was done, on 20 May 1786 Hutton signed an agreement with Joseph Johnson, publisher, to write 'a Mathematical & Philosophical Dictionary'. It was to be substantial – about eight hundred pages were anticipated – and Hutton was to be paid handsomely: around four hundred guineas depending on the exact length of the finished book (the sum was about twice his annual salary from Woolwich, although by this time that salary was noticeably low). Both parties apparently thought the book would be complete within four or five years, and Johnson would pay Hutton quarterly, twenty-five pounds each time: the balance to be made up after publication. Hutton undertook to supply drawings of the pictures and diagrams needed; Johnson agreed to supply Hutton with any and all books he needed to consult. On the whole it was a handsome and an optimistic document, indicating that both men thought a comprehensive dictionary of mathematics could be completed in a reasonable time and would sell well enough to be highly lucrative.

Beyond that, it is not perfectly clear what Hutton intended to achieve with his mathematical dictionary. The agreement named the *Encyclopaedia Britannica*, recently published at Edinburgh, as a standard of reference as far as size was concerned, and though it was not explicitly laid down as a model, there was surely something of that intention. Hutton's fondness for French culture and his admiration, expressed in print, for Diderot's *Encyclopédie* suggests he was looking across the Channel for inspiration for his own project, but the word 'Encyclopaedia' was not directly applied to the mathematical dictionary at this stage. Hutton would eventually, in his preface to the published *Dictionary*, point to the existence of a number of dictionaries on various subjects recently published and the absence among them of one devoted to mathematics or natural and experimental philosophy; he would rate the *Encyclopédie* 'stupendous' but suggest that it, as well as the *Britannica* and Chambers' *Cyclopaedia*, were alike deficient when it came to the details of particular sciences. A mathematical encyclopedia, then, rather than a mere dictionary, was perhaps in Hutton's mind.

We hear nothing more about the project for several years. Although Hutton would later say he worked on the book solidly for a decade, he was doing many other things as well, including, still, his teaching job at Woolwich. As work on the book progressed, though, Hutton witnessed from afar the demise of French enlightened culture as he had known it, and the transformation of France into (again) a declared enemy state. French literary and intellectual models became less acceptable in Britain, and the *Encyclopédie* could no longer be an acknowledged inspiration.

It also became obvious during that decade that Hutton could not possibly complete a mathematical encyclopedia on the scale he might have liked. He put it about that the article on algebra, which contained one of the most comprehensive and detailed histories of the subject yet written, had cost him two years of reading and writing. That is borne out by surviving manuscripts, which show Hutton making detailed notes and synopses of every old algebra book on which he could lay his hands, as well as translating *in extenso* contents pages, prefaces and even in one case a book-length section from the sixteenth-century Italian of Niccolò Tartaglia. A dictionary or encyclopedia conceived on such a scale would not have been completed in a lifetime.

So the project became, perhaps for a combination of reasons, less like an encyclopedia and more like the 'dictionary' of mathematics and science envisaged in the original agreement: in a sense somewhat like that of Johnson's 1755 *Dictionary*. Hutton wrote a large number of succinct dictionary-style definitions covering terms and concepts from mathematics and science, and also ranging across astrology, music and military theory. He took what amounted to the contents of a decent primer on arithmetic and chopped it into medium-length articles for insertion into the dictionary. He did the same for geometry, mensuration, astronomy and military mathematics. He added biographies of ancient and modern mathematicians, based on the one hand on their works or secondary sources, and on the other on Hutton's personal knowledge gained through correspondence and as editor of *The Ladies' Diary*.

Just how much of this was Hutton's own personal work is not certain. A letter from Nevil Maskelyne to Hutton survives from this period in which Maskelyne suggested some text for the article on Parisian astronomer Jérôme Lalande. Direct acknowledgement in the published text was rare, but Hutton mentioned for instance Abram Robertson, of Christ Church, Oxford and Dr Damen, Professor of Mathematics at Leiden as sources of information; surely there were many more. John Playfair was involved with sending Hutton information about the mathematical family of Gregory from Edinburgh, although not much of it found its way into the dictionary; Edward Waring of Oxford supplied ten pages on algebra.

Citations within the *Dictionary* as published make it clear that Hutton's main sources were printed, however. A list of them would amount to little less than a catalogue of Hutton's mathematical library, which contemporaries acknowledged was the best in the country. He possessed quite a number of early mathematical books: the algebra texts mentioned above, but also early editions of Greek geometrical works. He also had a wealth of British and continental material from the seventeenth and eighteenth centuries: hundreds of volumes in four or five languages. If Newton and Newtonians were conspicuously well represented, it was also made quite clear in the *Dictionary* that the author was a man who kept up to date with such publications as the *Proceedings* and *Memoirs* issuing from Paris and Petersburg: the output of 'Learned Societies throughout Europe', as he emphasised in the preface. The names of Leonhard Euler, Jérôme Lalande and Alexis Clairaut were frequent in the *Dictionary*, and Hutton took every opportunity to display the breadth of his mathematical gaze.

In a letter of 19 December 1793 Hutton informed Maskelyne that he had 'just got to the end of the Alphabet with my Dictionary', and that he had begun to send sheets to the press. The manuscript, he stated, was written mostly in his own hand, although teenage Charlotte did 'some copying'. Hutton was weary of the project, thanking God in this letter that it was 'out of my hands'. In a letter to Robert Harrison in Newcastle in 1794 he referred to it as 'an immense task'; in the preface to 'great labour and reading'. It took me a good thirty hours just to leaf through the *Dictionary*; the thought of fair-copying it using quill and ink is quite a horrible one.

The dictionary was indeed much larger than Hutton's agreement with Johnson had anticipated; all told, the manuscript filled more than fifteen hundred sides of folio paper. We do not know whether Hutton had discussed this burgeoning with Johnson, or how the publisher felt about the matter. He insisted on using a very small typeface for the book, over Hutton's objections, but even so the printed *Dictionary* filled around twice the number of sheets the contract had stipulated.

The reasons for the large size of the *Dictionary* were various. The long articles were in some cases very long, and they ranged across Hutton's personal enthusiasms from algebra to hot air balloons. The medium-sized articles amounted to four or five short books in themselves. The biographies were bulked in many cases by comprehensive lists of the subjects' works, and Hutton's interest in the history of his subject lengthened many of the other articles too, as did a tendency to give both sides of every disputed question. And the short definition-style articles were legion, covering a much wider range than might reasonably have been expected. The contract and the eventual title page both specified a 'mathematical *and philosophical* dictionary' and the preface pointed to a desire for greater comprehensiveness than had been achieved in any predecessor, for inclusiveness rather than its opposite in cases of doubt. But few readers would really have turned to a dictionary of mathematics in order to find out about *balconies* and *balustrades*, still less about *barricades*, *bastions* or *batteries*.

More than a year passed between Hutton sending the first sheets of the *Dictionary* to the press and the first parts going on sale. It was, like Hutton's previous large works the *Mensuration* and the *Miscellany*, initially issued in separate numbers, the first appearing on 31 January 1795. There were numerous advertisements in the press, and a single-leaf puff for the *Dictionary* also circulated, Hutton himself sending copies to some of his correspondents.

The separate numbers were gathered into four 'parts' and bound in two volumes, the whole going on sale in February 1796, three months later than the advertisements had promised. Something seems to have gone wrong with the preface, where Hutton might have expected to justify the size and shape of the book in some detail: it merely repeated text from one of the advertisements, and in some copies of the *Dictionary* it was bound not at the beginning of Part 1 but at the beginning of Part 4, in the middle of the second volume. The result was an oddly austere appearance for the first volume, which went straight from title page to the entries under A beginning with *abacus*, with no explanation of who was meant to use the dictionary, how or for what. There were also three dozen additional entries under the general title of 'addenda et corrigenda' at the end of Part 4; evidently the 'finished' product of late 1793 had already been the subject of Hutton's second thoughts during the subsequent months, and again the result was difficulty for the reader, who had no reason to expect that (parts of) the entries for *achromatic*, *Bernoulli* or *canal* would be found far out of their alphabetical sequence.

The *Dictionary* had generated a good deal of anticipation among Hutton's friends. John Playfair, in Edinburgh, had been asking after its progress since August 1792; among the Greenwich Observatory circle it was expected to be a 'capital performance', and several members of that group declared themselves purchasers shortly after the first parts went on sale, despite the fairly high price: a shilling a number, or two pounds fourteen shillings for the whole thing. In a letter of about 1795, before the second volume was published, Hutton referred to 'the very favourable manner' in which the *Dictionary* had been 'received by the public'. Another Scottish friend, John Leslie, had professed himself in October that year 'much entertained' by the early parts. Several reviewers – including, surely, some of Hutton's personal friends – judged the *Dictionary* learned, masterly and useful: 'a performance of immense erudition, and by which the fame of its author is fully established'.

But there were, not surprisingly, those who disagreed. For those determined to be hostile it was quite possible to see the *Dictionary*'s diversity of material, of styles of writing, as catastrophic unevenness. To see its range of subject matter as bewildering, to carp at its size, price, and the personal idiosyncrasy of many of its choices. More than one commentator disapproved of the whole project. One argued that a compilation of this kind was of use only to 'smatterers and would-be scholars', to '*dilettanti* and vain pretenders to philosophical reputation'. Another went so far as to charge Hutton with the 'prostitution of the best abilities and deepest knowledge' to the commercial book trade.

It didn't help that in the preface/advertisement Hutton boldly claimed that the *Dictionary* was 'of an equal and uniform nature and construction throughout': it was most certainly not. Under some headings it contained far more information than might have been expected, under others rather less; it contained much that no one would have expected it to contain at all. It was hard to predict whether any particular term would be included or not, or whether it would be treated briefly or extensively. Hutton also claimed the book was of 'moderate size and price' but in fact it was expensive and bulky by almost any standard.

For all that, the *Dictionary* was certainly a success rather than a failure. The balance of reviews was positive. A second edition appeared in due course, and imitations and derived works followed. Quotation from it and reference to it became a regular feature of English writing about the subjects it covered. As a monument to British mathematics it appears to have achieved all Hutton had hoped.

∞

His mind, however, had been on other things since the completion of the *Dictionary*; while it was going through the press he was struck from a direction he could never have expected, and with a severity from which he could never wholly recover. Late in 1794 his beloved youngest daughter Charlotte, quite suddenly and with little or no hint of earlier illness, ruptured a vessel in her lungs. She died on 24 September. She was sixteen.

Life and duty carried on, but Charles and Margaret were for a long period beyond consolation. Hutton's letters of this period show a man desolate almost beyond words, speaking of what was left of his life as a 'gloomy remainder'. Of all his children it was Charlotte who had most seemed to bear his intellectual gifts, to whom he was closest, on whom he most doted and in whom he most hoped. 'Never did I think it possible I could feel such severe affliction', he wrote to a friend. Five years later, in 1799, a printed account of Hutton's life mentioned that 'he has never ceased to lament the loss . . . [has] never quite recovered his wonted spirits and liveliness.'

There was a curious coda to Charlotte's death. Just a few days earlier she had been awake before the rest of the family, and when they found her in the parlour she told them a remarkable dream she had had. Her written account of it was printed with the notice of her death in *The Gentleman's Magazine*, and so it is that we have a few lines of prose from a young woman whose voice would otherwise be silent.

I dreamt that I was dead, and that my soul had ascended into one of the stars; there I found several persons whom I had formerly known, and among them some of the nuns whom I was particularly attached to when in France. They told me, when they received me, that they were glad to see me, but hoped I should not stay with them long, the place being a kind of purgatory, and that all the stars were for the reception of different people's souls, a different star being allotted for every kind of bad temper and vice; all the sharp tempers went to one star, the sulky to another, the peevish to another, and so on. Every body in each star being of the same temper, no one would give up to another, and there was nothing but dissension and quarrels among them. Some of those who received me, taking offence at the information my friends were giving to me a child, it made a quarrel, which at length became so rude and noisy, that it awaked me.

∞

And as their grief was losing its keenest edge, the Huttons were struck again.

Camilla had lodged in Barbados while the British took Martinique, St Louis and Guadeloupe from the French in 1794. But those early gains were followed by an outbreak of yellow fever that cost far more lives than the fighting. Thousands perished among the British forces, and in the winter of 1794–5 – just a few months after Charlotte's death – Hutton learned that Camilla and her husband and child had perished with them at Guadeloupe.

The life of a soldier's wife had its perils, but this second loss was no less hard for all that. Hutton received initially only rather confused reports, which said in addition that his son Henry, now a captain in the Royal Artillery and also on the West Indian station, had lost an eye and was a prisoner of war in Martinique. Hutton had long had some sympathy with radical causes and some distaste for the Pitt and Pittite ministries; little wonder that in February one of the assistants at the Greenwich Observatory, David Kinnebrook, found him 'very severe upon Mr. Pitt & Administration in general; for carrying on the War'. Kinnebrook added dryly that Hutton 'has reason to complain'.

Fate had another surprise in store, however. More than six months later, in early August 1795, Henry Hutton arrived at Woolwich. How much warning he gave we don't know; he had indeed lost an eye, and he was indeed a prisoner on parole from the French. He was accompanied, in a scene worthy of grand opera, by an Irish nursemaid and a two-year-old boy.

Closer to the scene of the action than his father, Henry had heard the truer word; infant Charles Vignoles had *not* died with his mother and father. Henry received permission from his commanding officer to go to Guadeloupe, now held by the French again, intending to cross the lines under a flag of truce. When he got there he found the remaining English garrison in such a desperate state he felt compelled to join them, and in the subsequent fighting he was wounded and taken prisoner. The French commander gave him permission, nevertheless, to continue his mission of rescue, and he was able to locate Charles Blacker Vignoles in the home of a French merchant named Courtois who had taken in, and buried, his parents. Henry returned home on parole with the child and his nurse.

It was a story that would have strained belief, had young Vignoles not been accompanied by a piece of paper on which his dying mother had scrawled the names and addresses of his uncle and grandfather. Courtois's own letter, written on the same sheet, told an agonising tale. Charles and Camilla had been reunited at Pointe-à-Pitre on Guadeloupe during 1794, but as the English succumbed to fever the French retook the island and the town was besieged.

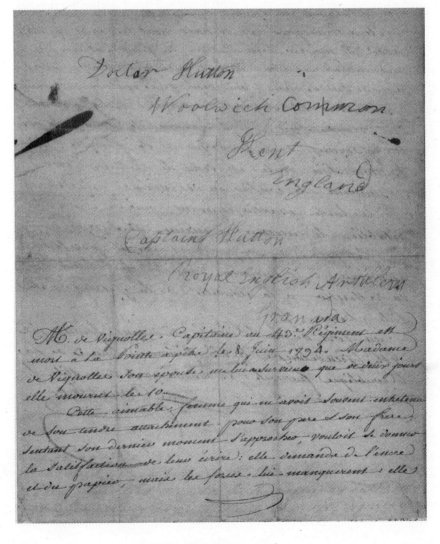

Camilla's dying scrawl.

During the chaos Charles took fever and died in his lodgings on 8 June, followed two days later by his exhausted wife. 'I had them given a decent burial'. The child, too, was infected, but he survived; Camilla had wished her own family to take him in, and he was entrusted to his uncle Henry two months later.

Charles Hutton was fifty-eight, and the addition of a two-year-old child to his household was hardly an event he would have countenanced in less extraordinary circumstances. But Camilla's dying scrawl, and Courtois's letter which spoke most painfully of her regrets, left him little choice. So the Cube House once again acquired a nursery and schoolroom, and Hutton added to his roles that of Vignoles's guardian, and to his worries his education and upbringing.

It was in this context, remarkably, that Hutton systematised his mathematical syllabus in full and in writing. He had continued to refine his presentation of mathematical topics to students, of course, through his long years of teaching at the Royal Military Academy. His *Guide* was now in its twelfth edition, the *Mensuration* and *Measurer* each in their third. More and more he found it frustrating to teach the cadets from a range of different books, using a section here and a section there as their training required, and he determined – as a number of his colleagues and predecessors at Woolwich had done – to write his own comprehensive textbook, so that everything he needed would be in one place, with no extraneous matter and nothing in the wrong order. It would also be the culminating, most complete demonstration of his skills as a pedagogue.

Hutton's *Course of Mathematics* was at first a manuscript compilation, as the *Conics* had once been, though the scope and duration of the Woolwich mathematics course meant that it filled two fat volumes totalling 750 pages. It was a work of synthesis, drawing together teaching material that Hutton found in the textbooks of his various predecessors and the mass of mathematics books produced in Georgian Britain, as well as mining his own earlier works for suitable material. In a way it was little more than a scissors-and-paste exercise, with a presentation of arithmetic rewritten on the model of the *Guide* and a presentation of geometry and practical geometry rewritten from the *Mensuration*, the *Measurer* and the *Conics*. Scientific matters such as hydrostatics had already appeared in the second half of the *Conics* and were in part taken over from there. But there was also much that was new; arithmetic was followed by a presentation of algebra, something Hutton had never written about in his earlier textbooks, and the basics of geometry were handled, newly for Hutton, in the theorem-and-proof style of his *Conics*. Calculus was included, as it was not in his earlier, more elementary works. In fact there was hardly a section longer than a paragraph copied from Hutton's other books without some modification.

Tried out with the students in its manuscript form, the course of mathematics found some favour, and Hutton approached the management of the Academy to fund its publication. He managed to persuade them that this key textbook by a key member of staff should be quite handsomely supported, and the Board of Ordnance undertook to pay a hundred guineas for the first three hundred copies, and seven shillings per copy for as many more as were needed. In the event the sums were increased, Hutton finding he had underestimated the cost of printing the books.

The first copies issued from the press in 1798, and Hutton's *Course of Mathematics* became a staple of life at the Royal Military Academy. Every cadet had to buy a cheap copy; every graduating lieutenant was given a bound copy as a present. It was fairly clear from the start that the book would outlast Hutton at the Academy; in fact it remained in use for decades across three continents. It was a fine monument to Hutton's style of teaching and to all he had achieved at Woolwich.

But there were undoubtedly flaws. Despite the rearrangement and rewriting the *Course* remained in large stretches merely a compendium of what Hutton had written elsewhere, and not all the changes, compressions and adaptations to the military point of view were improvements. If complexity was trimmed, say from the long series of theorems on conic sections, so were long and vital series of example problems, and in places such as the arithmetic the substitution of problems dealing with specifically military matters felt forced ('Suppose 471 men are formed into ranks of 3 deep, what is the number in each rank?'). Volume 2 in particular felt like a mere miscellany of different topics: trigonometry and mensuration; conics; dynamics; hydrostatics and hydraulics; fluxions and their mathematical applications. And there was an unevenness of tone between different parts of the book, betraying their origins in textbooks written for rather different constituencies at different times in Hutton's life.

It received mixed reviews. Too hard, said some; too compressed, others. Much was good but perhaps too little was new or was properly (re)considered. One reviewer was 'not perfectly pleased' with the order of subjects and thought some difficulties 'not sufficiently softened for beginners'; another judged it a mere selection from other works. Cadets at the sister college at Addiscombe found the *Course* as indigestible as their dinners:

> How shall I name the o'erbaked ribs of beef,
>   The stringy veal, and greasy legs of mutton,
> Whose very sight oppress the soul with grief?
>   (Almost as great as brought on by a Hutton.)

∞

The *Dictionary* and the *Course* were large projects, the *Dictionary* a hugely laborious one. But they had exhausted neither Hutton's energy nor his desire to place mathematics before the reading public on the largest possible scale. His research for the *Dictionary* had made him more aware than ever of the volume of interesting mathematics and writing about mathematics in circulation on the European continent, and one name prominent in both popularisation and mathematical history was that of Jean-Étienne Montucla. He had published a history of attempts to square the circle in 1754 and a comprehensive history of mathematics in 1758; Hutton greatly admired both works, and there is some evidence he may have corresponded with Montucla about revisions to the latter. And in 1778, with a second edition in 1790–1, Montucla produced a new version of an old favourite, the 'Mathematical Recreations' of one Jacques Ozanam. First published in 1694, this was a compendium of games, puzzles, riddles, tricks and mathematical popularisation. The book had been translated into English early in the eighteenth century, but Montucla's new French version was massively expanded and updated, and Hutton saw a chance to put a great deal of well-written mathematical exposition in front of an English audience without having to compile it from scratch himself. The four volumes of 'Montucla's Recreations' duly appeared, under the title *Recreations in Mathematics and Natural Philosophy*, in English translation in 1803.

It was a jolly compilation, ranging across arithmetic, geometry, and many kinds of applied mathematics including astronomy, navigation and music. There were counting games and card tricks, optical illusions and instructions for making fireworks. How to draw a circle without compasses; how to construct a magic square; how to build a water clock; how to build a microscope.

The title page and preface stated that the book was – compared with its French parent – much augmented, enlarged and improved with 'notes, remarks and dissertations'. But the careful reader would search many chapters in vain for any such items. There were a few obvious changes such as a table of English rather than French weights and measures, or a list of eclipses visible from London instead of Paris. There was an occasional simplification to the prose, by the removal of a metaphor, say. Now and again there was a mistake of Montucla's to correct, a facetious footnote to omit or an algebraic demonstration to add. Sometimes changes were marked 'editor' but more often they were not. Mostly, though, it was a very straightforward translation. Footnotes remained footnotes; remarks remained remarks. There was very little in the hundreds of pages of 'Hutton's Recreations' that was not in Montucla's.

It was, indeed, never quite declared in the prefatory matter that Hutton was the translator, though it was everywhere implied. If there were added a few sections about some of his favourite hobbyhorses – the specific gravity of various substances, steam engines, hot air balloons – there were also at least a few places in the *Recreations* which it is hard to believe Hutton worked on: notably a remark about the impossibility of determining the masses of the planets which stands in flat contradiction to his own published determination of those masses. The translation itself occasionally reads like the work of a man less than fully conversant with mathematical terminology ('squarrable'; 'cube foot') and the references to recent – particularly British – work, that Hutton would easily have been able to add, are altogether absent.

Did Hutton translate the *Recreations*? Or did he do no more than lend his name, cast an eye over, perhaps add a few notes here and there about subjects that interested him, to a volume translated by one or more invisible assistants? Was this, as the harshest critic said of the *Dictionary*, merely a project cooked up in collusion between the booksellers and a big name? It's hard to say. Hutton probably didn't do all the translating work, but just how big or small his contribution was we are unlikely ever to know.

In a similar vein, it was and is sometimes said that Hutton translated a second work on popular mathematics during this period: *Select Amusements in Philosophy and Mathematics; proper for agreeably exercising the minds of youth*, by one M.L. Despiau. In fact he merely provided an endorsement for the English version. But the title page was set out in a way that made it ambiguous just how much was 'by Professor Hutton', and some were evidently fooled. Hutton's was becoming a name about which mathematical publications accreted, whether relevantly or not.

Reviewers on the whole liked the *Recreations* project. *The British Critic* found the book not merely entertaining but useful, and gave it 'warm commendation':

> It is no small proof of a genuine regard for philosophy, when men, whose peculiar privilege it is to move in the most exalted sphere of science, will condescend to smooth the rugged path to eminence, and strew with flowers the wearisome way to the Temple of Knowledge.

The Monthly Review concurred, reckoning the book 'well adapted to lie on the table or chairs of our parlours', since science would not lose its dignity by losing its formality. Both reviewers approved the project of popularisation in general and thought this a fine specimen of the genre. *The British Critic* was similarly enthusiastic, singling out various sections for praise and quotation: 'The whole work indeed reflects much credit upon a man, who has already deserved so well of the scientific world,' though he excepted the chemical section which he thought poor. He thought it would make a useful form of recreation in schools.

A dissenting voice came from *The Imperial Review*: alas we don't know whose voice it was. The reviewer implied that the *Recreations* contained much matter that was 'insipid and puerile' and that Hutton had done far too little to the book in passing it through his hands, making only trifling additions and alterations: indulging in negligence rather than adding the thorough updates and pointers to recent literature one would have expected. The reviewer, like others, suspected the work might have been Hutton's in name only.

∞

Almost incredibly, Hutton had in hand yet another project, and one on an even more heroic scale.

The *Philosophical Transactions* of the Royal Society now ran to nearly a hundred volumes, and even such a devoted and assiduous book collector as Charles Hutton could not get hold of them all; some were scarcely to be had for money. He knew from personal experience what a chore it therefore was to chase a subject through multiple volumes that might have to be consulted in different libraries. There had been printed compilations, 'abridgements of the *Transactions*' before, but none – at least in English – for some decades. An attempt was announced in 1802, but flopped and was quickly abandoned. The publishing firm of C. and R. Baldwin jumped into the gap, and in March 1803 announced in a short prospectus that they were going to do the thing properly. They had engaged a team of three: Charles Hutton of Woolwich, George Shaw of the British Museum and Richard Pearson of Bloomsbury Square. Under their care there would appear, in weekly numbers from June that year, a new abridgement that would serve the function of a complete set of the *Transactions* for those who didn't have them and couldn't obtain them, containing in full the most interesting articles and in brief the moderately interesting ones, while leaving out a few of the wholly superseded or uninteresting. The whole would be in English (except where 'from the peculiarly delicate nature of the subject, there would have been a manifest impropriety in giving them in English') and enriched by notes and biographies. By 1806 eight volumes had appeared and in 1809 the *Abridgement* went on sale complete, in eighteen volumes totalling something like fourteen thousand pages.

The project had much in common with the 1775 *Miscellany* that brought together lightly edited reprints of *The Ladies' Diary* from the previous seventy years. It was widely reported that the *Abridgement* was under Hutton's principal care, and it would be repeatedly stated in print that he was paid the fantastic sum of six thousand pounds for his trouble. If that colossal figure is anything like the truth it can only have been envisaged that he would use it to employ assistants to do much of the work: translating, editing and annotating. We hardly need to imagine Hutton, in his late sixties, actually penning the eighteen volumes or any significant fraction of them. Nevertheless, there were places that bore his distinctive mark. Annotations on papers of personal interest to him – such as certain items to do with the density of the earth – were occasionally voluminous, and neither did he hesitate to do some discreet rewriting of his own paper on that particular subject.

One of the most notable achievements of the *Abridgement* was as a subject classification of the matter that had appeared over the 136 years of the *Philosophical Transactions*. From volume 3 (following criticism, in fact, of volumes 1 and 2) a very full classification was adopted in the contents list, and it featured mathematics and 'mechanical philosophy' (much of which was applied mathematics) as the first two of its eight main headings. This was, as well as anything else, a substantial assertion about the place of mathematics in the intellectual world and in the work of the Royal Society. (A 1787–91 Paris abridgement of the *Transactions*, in fourteen volumes arranged by subject, did not have a volume devoted specifically to mathematics.) Under pure mathematics alone there were nearly three hundred articles (a quick look at the contents pages would have revealed that the number fell off markedly during the period of Joseph Banks's presidency).

This was, of course, a bold, even a cheeky project as well as a huge one. Just before Baldwin's announcement, Hutton wrote to

Banks 'signifying his intention to undertake the care of arranging and printing a new abridgement of the Philos: Transactions, and trusting that the President and Council would please to countenance this undertaking'. Council noted the matter and gave it no further discussion, but by 1809 permission had nevertheless been received to dedicate the work to 'the Right Honourable the President, the Vice Presidents, the Council and the rest of the Fellows of the Royal Society of London'. It's possible that the abandonment of a planned supplementary volume containing a history of the Society and additional biographies of Fellows – which would surely have been an occasion for some Banks-bashing – was connected with this permission. If the project gave anyone the impression of an improvement in relations it was a deceptive one, unfortunately. But whatever Banks really thought, Hutton had succeeded in putting his own mark on how a large fraction of readers for the foreseeable future would view the Royal Society and its activities.

There were criticisms from *The Monthly Review*, which was becoming quite consistently hostile to Hutton and his work. There the reviewer thought a full republication of the *Transactions* would have been better, or failing that perhaps reduction into the form of a compendium or encyclopedia; he found the selection bewildering and even capricious. But the *Abridgement* was well received by most, its balance and even-handedness coming in for particular praise. *The British Critic* called it a work 'even of national impor-tance', possessing 'spirit, taste, and judgment' and, judging from the early volumes, shaping up to be 'a national honour'. *The Literary Journal* praised its accurate, 'perspicuous and unaffected style'. One of Hutton's friends, Thomas Leybourn, said the early volumes 'have fully justified the public expectation, a circumstance which will not appear remarkable when it is recollected that the execution of the work is confided to men of the highest eminence in the departments of Science they have engaged to superintend'. There was even a favourable notice in France. Of Hutton's large-scale projects from the second half of his career the *Abridgement* was arguably the most successful, and most unambiguously achieved his oft-stated goal of public utility.

∞

During the long decade from the appearance of the *Dictionary* to that of the *Abridgement*, Hutton's activity and his public persona were focused no longer on practical experiment, hands-on surveying and the other dirty business of practical mathematics, but on text, and text in huge quantities. Whatever the role of assistants, his work on *Dictionary* and *Course*, *Recreations* and *Abridgement* must have taken up a substantial portion of his time, against a background of personal disaster and the chaos of war at the Royal Military Academy.

It was possible to be very sharply critical, to say that Hutton had spent a decade and more on scissors-and-paste work. But that was to miss the point. The compilations and translations he had produced were not about intellectual innovation or conceptual novelty. Rather, he had carefully, deliberately, made himself the leading voice speaking for mathematics in English. And he had given the English-speaking world a series of massive monuments to its mathematical culture: monuments whose endurance was never in serious doubt.

More than that, these works were finely crafted tools. Citations of 'Hutton's Dictionary', 'Hutton's Recreations' and the rest would become ubiquitous in discussions of scientific and mathematical subjects, in writing about mathematics and its place in culture and education. It's not too much to say that Hutton, by providing them, changed the way English speakers spoke and thought about mathematics. His view of what mathematics was, of what it included and excluded, and where and how it was relevant, became the norm, and he did more than anyone else to make mathematics an accepted part of British culture, an accepted strand of British science.

Yes, much of the work on these books could have been done by any competent translator, editor, compiler: and some of it probably was. But Hutton knew what he was about. In the face of hostility to mathematics from parts of the scientific and the literary worlds, through these massive works he both consolidated his own reputation beyond doubt, and gave unique, permanently valuable service to the mathematical culture he knew and loved.

10

Securing a Legacy

Mount Schiehallion, in Scotland: June 1801. Two thousand feet and more above sea level; long views over the Grampians away from the ridge. Steep, rugged, 'a very beautiful conoidal shape'. Charles Hutton's friend John Playfair, mathematician and geologist, treads the high ground. Retreads the ground, indeed, that Maskelyne and Burrow trod back in the 1770s.

All that remain from 1774 are the ruins of the two observatories, and two cairns on top of the mountain. The long chain of surveying stations, meticulously positioned and measured, is gone without trace.

Playfair nevertheless tries to follow the lines of the original survey, using theodolite, sextant and compass to determine the lines of the different strata. The strata, nearly vertical, break out through the surface, and Playfair, hammer in hand, collects specimens of each, notes their positions. Quartz, lustrous as enamel. Schist and limestone; porphyry and greenstein. Hutton's paper from the 1778 *Transactions* is seldom out of his hand.

∞

The *Course*, for all its imperfections, was the culmination of Hutton's work on the wide-ranging mathematics syllabus at Woolwich. After it was done, wartime conditions left him little opportunity to develop his teaching further; he was becoming more and more a manager of other men's work and of a complex system of academies and classes, with their attendant tests and documentation.

The Academy at last moved to new buildings on Woolwich Common in 1806 in order to relieve what had become intolerable overcrowding on the riverside site. Not all the cadets could be accommodated even then; only the senior academy of 128 went to the new buildings. The remainder, now numbering 180, were split between the old Woolwich Arsenal site and the Royal Military College at Great Marlow. But the new buildings were a great improvement; each of the four classes now had its own separate housing and classroom. There were a library, a lecture room and two model rooms; a fencing room and two racquet courts; assembly areas and gardens.

The governor, the inspectors and the professor of fortification were now provided with houses outside the Academy precincts; others found their own accommodation. The Cube House was purchased from Hutton and became an infirmary. But there was then an unexpected embarrassment about the village of houses he had built beside it. During a visit to the new Academy the King noticed that the view from it down towards the barracks was blocked by 'some new houses'. 'Let the whole be purchased, and the obstructing parts removed,' decreed the monarch. Hutton agreed to accept a valuation determined by a surveyor from the Board of Ordnance, and their Mr Wyatt settled on twenty years' worth of the total rents Hutton was receiving. Six of the houses were demolished at once, though the rest remained; precisely where Hutton and his family now lived is not clear.

On top of this, Hutton was tiring of his work at the Academy. He was in his thirty-third year there, and with no prospect of peace in Europe the disruption of war was perhaps becoming hard to bear. Just a year later, in June 1807, he was granted permission to retire, with a pension of five hundred pounds per year. The sum was large enough to be a compliment; more, indeed, than he had been paid as salary for much of his time at Woolwich. He agreed to stay on as examiner of both cadets and new masters, and to provide advice when it was needed.

Rather than stay near the Academy, though, he moved into London, close to the Inns of Court where he had long kept rooms. His new town house on Bedford Row was near both the British Museum and Somerset House, where the Royal Society met (if that still mattered): close to London's intellectual centre, and to some of his friends.

He was affluent; the sale of his land and houses on Woolwich Common had left him with what contemporaries reckoned a small fortune, perhaps forty thousand pounds: equivalent to well over a hundred years of his salary at Woolwich. Some he invested in a bridge-building project that failed to pay any dividend for many years, but the rest brought income at something like 5 per cent. Together with his pension, then, he had around fifteen hundred pounds a year, at a time when Jane Austen's Dashwoods were making do with a third of that in their cottage on the Barton estate.

It was a pleasant house, the street broad, light and airy. A regular round of visits and visitors soon formed, and the house was crammed with books. The Huttons had two pianos; Isabella certainly played, and perhaps Margaret too. Isabella dearly loved a play, and Hutton appears to have been fond of the theatre as well. In 1809, rising prices at Covent Garden Theatre led to riots, and a committee formed to raise subscriptions and build a new theatre. Hutton sat on the committee.

Hutton was seventy at his retirement; his own *Dictionary* contained the official table of life insurance premiums, and it only went up to the age of sixty-seven. One of the activities to which he devoted himself more and more assiduously during his last decade at Woolwich and in his retirement was the securing of his intellectual legacy.

This he did first through people. At the pinnacle of the philo-math network, the retired holder of a key role at a highly visible institution of national importance, still a Fellow of the Royal Society even if he no longer attended its meetings, and a Fellow of three foreign societies, Hutton had often been asked to recommend mathematicians for teaching jobs and other work. He plied this trade well, and any number of provincial mathematicians were plucked at his word from the ranks of *The Ladies' Diary* and installed in positions of prestige and responsibility. Lewis Evans, a country schoolmaster who got to know Hutton through the *Diary*, became a mathematical master at Woolwich, though he had to be asked more than once. David Kinnebrook, another *Diary* contributor, became an assistant at the Greenwich Observatory. Charles Wildbore, a contributor to *The Gentleman's Diary*, became its editor. Edward Riddle moved from private teaching in Newcastle to become master of the Trinity House School, Newcastle, and then master of the upper mathematical school at the Royal Naval Hospital, Greenwich. John Bonnycastle, yet another *Diary* man, became Hutton's first assistant at the Royal Military Academy. And it was not only *Diary* contributors who benefited. Margaret Bryan, a schoolmistress, received a testimonial and crucial encouragement for her book about astronomy. Major Edward Williams and Lieutenant William Mudge of the Royal Artillery were recommended to conduct the Trigonometrical Survey of Great Britain; Mudge (the same Mudge whom Hutton had once turned back into the lower academy until his mathematics should improve) later became governor of the Royal Military Academy. Sir John Leslie was recommended for the professorship of mathematics at St Andrew's, and later for that at Edinburgh;

the young Charles Babbage was recommended (unsuccessfully) for mathematical professor at the East India Company's college in Hertfordshire.

Hutton felt a sense of duty about all this: 'serving and encouraging very able and worthy persons, and . . . supplying useful institutions with good and proper teachers', as he put it. But, like much that he did, it also drew some attention to himself, built his own reputation; and the more he did in this kind, the more he was able to do.

All these men and women, in turn, owed Hutton something and could be relied on to do at least a little to promote his interests and serve his legacy: for instance by praising and promoting his books and disparaging their rivals. Such things mattered. Back in 1788 there had been a quite extraordinary fracas when a jealous rival of schoolteacher Lewis Evans put it about that Evans had 'in public company' dispraised Hutton's *Guide*. Evans not only wrote to Hutton in the greatest haste to deny the charge; he also organised a joint letter from the governors of his school testifying to the same effect. Patronage worked both ways, and damaging Hutton's reputation – or being thought to have done so – could have been costly indeed for Evans.

The most important promoter of Hutton and his work was a bookseller and mathematician who rejoiced in the name of Olinthus Gregory (his Christian name is the Greek word for a kind of wild fig). Forty years younger than Hutton, he hailed from Yaxley in Huntingdonshire and manifested early interests in both theology and mathematics. Unwilling to subscribe to articles of religion, he could not matriculate at an English university. But by 1798 he was living in Cambridge, working as a bookseller and probably doing some private teaching of mathematics. Rumour had it that his unpublished treatise on the slide rule impressed Hutton; he also proposed and answered questions in *The Ladies' Diary*.

Olinthus Gregory.

When Charles Wildbore died in 1802, Hutton recommended Gregory to the Stationers' Company as his successor, making him editor of *The Gentleman's Diary* and probably also of the successful comic almanac *Old Poor Robin*. In the same year he became second assistant mathematics master at the Royal Military Academy. Thereafter he was beyond doubt the crown prince to Hutton's mathematical kingdom. He translated papers for the *Abridgement*. Together with William Mudge he did work with ballistic pendulums at Woolwich in 1815–17, taking some of Hutton's results further and promoting their adoption in practice. He was involved in testing a proposal to reduce the windage of cannon, which arose directly from Hutton's own ballistics work and suggestions.

When Hutton at last retired from his own work on the almanacs in 1818, Gregory took on both *The Ladies' Diary* and the role of general superintendent of almanacs. Like Hutton, he used the position to dispense mathematical patronage; like his mentor, too, he was admired for the use he made of the *Diary* to promote mathematics and educate the talented young.

Gregory also continued with his own independent work: a trig-onometry primer in 1816 and a dissertation on weights and measures published the same year; the ambitious, eccentric *Pantologia*, a twelve-volume dictionary and encyclopedia of 'Human Genius, Learning, and Industry', in 1808–13. In 1823 he would determine the velocity of sound experimentally. Meanwhile he continued to write on theology. His career culminated with his role in the foundation of the non-denominational London University in 1827. He became professor of mathematics at Woolwich in 1821, and in due course the Stationers' Company awarded him a handsome pension for his work on the almanacs. Charles Knight, nineteenth-century promoter of learning reform, gave him an entry in his biographical study of great men.

It was a trajectory with which Hutton could of course identify, and he could take justified pride in the fact that Gregory owed much of his initial rise to his own help. The Academy, the almanacs and even Hutton's publications were safe in his hands.

∞

More could be done, though, and more directly than merely appointing a sort of intellectual heir and executor. Hutton also worked to secure his legacy through his publications, and a significant proportion of his time was taken up with preparing edition after edition of each of them, and seeing those new editions through the press.

The *Tables*, the *Mensuration*, and even his first book the *Guide* were all proving long-lived, and many a school relied on the last two for instruction. Indeed, unusually for an author on technical subjects, Hutton enjoyed a significant income from the sale of his books: he referred with some pleasure to the 'liberal encouragement of the Public' as a source of the means he now enjoyed. The *Guide* was 'held in high estimation' and was named in many a school curriculum around the British Isles in the early nineteenth century. The *Mensuration* was 'the most complete work on the subject ever published' according to the Edinburgh *Annual Register*, and remained a standard reference in the classroom as well as elsewhere. The *Tables*, too, had established themselves as such a standard that when one of Hutton's correspondents reported on the forthcoming appearance of a rival work he suggested it would be little more than 'an inaccurate transcript' of Hutton's.

And it was not only in Britain that his works were valued. Hutton was living proof that the British mathematical tradition was viable as an export product. A version of his *Guide* was one of the first accounts of bookkeeping to be printed in America (at Philadelphia, in 1788), and there were many subsequent American editions, variously transformed. His *Course* was adopted at West Point Military Academy from its opening in 1801, its contents becoming the basis for the mathematical course taught there. Robert Adrain, professor of mathematics at that academy, thought the *Course* 'one of the best systems of mathematics in the English language', and in 1812 he published an American version, with notes and other changes, condensing, reorganising and correcting.

West Point also had a copy of the *Dictionary* in its library, and that book, too, received the compliment of an American edition – of sorts. In 1817 one Nathan S. Read published *An Astronomical Dictionary* whose title page admitted with disarming frankness that it was 'compiled from Hutton's Mathematical and philosophical dictionary'.

If the English-speaking world appeared to have been conquered by Hutton's work, what of that harder market to crack, the French? Here the story is a more explosive one, since France was an enemy state from 1793. Just before that date, an unknown translator made a French version of Tract IX from the 1786 *Tracts*, Hutton's second published account of his ballistics work. Apparently it was done at the request of the chemist Guyton de Morveau. A story also circulated – even if untrue, the fact that it was plausible speaks volumes – that the great mathematician and astronomer Joseph-Louis Lagrange was working in Paris on Hutton's writings about ballistics. Report said that Lagrange, born in Turin, had been under threat of deportation in 1793, but was allowed to stay, again at the request of Guyton (who was a member of the Committee of Public Safety as well as a chemist), because he was doing work of national importance. What work was that? He was 'examining Hutton's Treatise on Artillery'.

Neither the translation nor Lagrange's response to the ballistics work was published. But in 1802, while the material remained unpublished in English, a French colonel named Villantroys got hold of the first part of Hutton's final discussion of his ballistics work and translated and published it in Paris. What the background was to this, we don't know; Hutton disapproved of the conduct of the war, but it seems incredible that he, an employee of the Board of Ordnance, could have deliberately supplied unpublished research on artillery to an enemy state. (Yes, for a few months in 1802–3 France was *not* at war with Britain, but Hutton was no more naive than the next person about the wisdom or the longevity of the Peace of Amiens.) Did Villantroys steal the text or have it stolen? I wish I knew.

Later still, printed works by French experts on artillery began to quote from and use Hutton's results. Charles Dupin included an illustration of the ballistic pendulum and Hutton's improved eprouvette in his 1820 *Military Force of Great Britain*; Antoine-Marie Augoyat quoted Hutton's work the following year in his

Mémoire sur l'effet des feux verticaux. Jean-Louis Lombard's translation of Benjamin Robins's work on gunnery contained a summary of Hutton's 1778 results. A French reviewer called Hutton 'as commendable for his observational talents as for the scientific methods he has used' to analyse the results, praising the thoroughness and consistency of his work. An English reviewer remarked, in turn, on how the French 'so eagerly possess themselves of every essay, investigation, and experiment of Dr. Hutton on the subject, as soon as it is made public'.

Both the American and the French uses of Hutton's work, gratifying though they were, bordered on piracy; editions in New York or Pennsylvania were most unlikely to have Hutton's explicit approval (or that of his British publishers), and the French version of the ballistics work he could not have avowed and may not even have known about. In part, Hutton's work was simply being absorbed into the common stock of knowledge. But there was such a thing as decency, and when, for instance, one Alexander Ingram in 1796 'corrected and enlarged' the *Guide* he was most likely taking gross advantage of Britain's lack of effective copyright protection. The same could be said of the apparently unauthorised *Keys* to the *Course* concocted by Daniel Dowling in 1818 and to the *Measurer* by J.M. Edney in 1824.

If Hutton could do little or nothing about the activities of American and French editors, printers and spin-off authors, he could certainly address the question of home-grown piracy by keeping up his own steady stream of re-editions of his most popular works. By the time of his retirement in 1807 there had been fourteen authorised editions of the *Guide*, twelve of the *Mensuration* and the *Measurer*, four of the *Tables* and five of the *Course*.

In order to keep readers buying these new versions, and keep them away from the unauthorised rivals or the work of his colleagues, Hutton was careful to make small but significant changes every time. Hardly an edition of his was without a few changed examples, improved explanations, minor reorganisation or rewriting, or insertions of new matter. The books, indeed, took on something of the character of permanent works-in-progress, the *Course* in particular functioning as a site to try out and announce new ideas, for instance in the theory of materials, the subject of ongoing experimental work at Woolwich. The *Mensuration* eventually became so altered and enlarged with new definitions and new figures as to be 'almost a new work'. Tables grew larger; examples were re-computed. Hutton welcomed suggestions from readers – of the *Guide* he had said in 1786 that 'Any hints of improvements that may be made, either in the Arithmetic or the Key, and addressed either to the author or publishers, will be attended to in the future editions of them.'

The same was at least implicitly true of all his books. From time to time *Diary* contributors, in particular, would offer suggestions for improvements to Hutton's textbooks, correcting errors or modestly proposing an improved example or a changed order of presentation. To the *Recreations* were added titbits from readers including a long and eccentric correspondence about divining rods, interesting less for its content than for the fact that the enthusiast who wrote to Hutton on the subject was Lady Milbanke, afterwards Lady Noel, Byron's mother-in-law and grandmother of Ada Lovelace. The *Dictionary*, too, was a particularly natural site in which to collect new material from readers and elsewhere, to reorganise and perfect, with in this case a still more obvious eye to Hutton's intellectual legacy, his stewardship of a particular world of mathematical culture and achievement.

Biggest of all was naturally the contribution of Olinthus Gregory. In 1811 a third volume of the *Course* was added to the original two, and although Hutton's name was on the title page it was acknowledged in a preface that Gregory had collaborated on the volume with his mentor and friend. The new volume dealt with conic sections, calculus, trigonometry, the solution of equations and various practical subjects including methods of surveying, the effects of machines and the pressure of earth and fluids. There was a summary of Hutton's work on ballistics as well as his customary concluding selection of questions. The intention was to deal with recent changes and improvements in the curriculum at Woolwich; but, as even Hutton tacitly admitted, the effect was somewhat miscellaneous. It tended to make the *Course* as a whole harder rather than easier to use in teaching, since much of what was in volume 3 really needed to be split up and inserted within volumes 1 and 2: it 'may best be used in tuition by a kind of mutual incorporation of its contents with those of the second volume', he admitted. The *Course* would have to wait several more years before posthumous editions effected the reorganisation and rationalisation it needed.

∞

About one particular part of his printed legacy Hutton became perhaps disproportionately obsessed. His work on the density of the earth, he became convinced, was in danger of being neglected, overshadowed, forgotten because of what had happened since. What had happened was partly his row with the Royal Society and the tendency of mathematicians generally to become invisible there, denied fellowship or publication. It had also happened that Banks's friend and supporter Henry Cavendish had devised a new, wholly different way to measure the strength of gravitational attraction and hence the density and mass of the earth.

Cavendish's was a beautiful experiment; he was, indeed, one of the most talented British natural philosophers of his generation. It used a torsion balance: a horizontal rod with weights at the ends, suspended by a fine thread and capable therefore of rotating horizontally. The force needed to rotate it was tiny, but could be measured, and the idea of Cavendish's experiment was simply to use the gravitational attraction of a ball of lead to attract the rod, to make it turn slightly away from its position of rest. Knowing the characteristics of the system (in particular, the period with which the rod oscillated once displaced) you could work out accurately the force the lead ball had exerted, and thus the strength of gravity. And hence by comparing the much larger force exerted by the earth on rod, lead and everything else, you could deduce the mass (or the density) of the earth.

As an experiment it had the advantage of real simplicity, and there was no need to go to a damp mountain for half a year to do it. It was a delicate experiment, but Cavendish obtained what he thought were reliable results and published them in the *Philosophical Transactions* in 1798. They flatly contradicted Hutton's value for the density of the earth; Cavendish made it about five and a half times that of water; Hutton more like four and a half.

Hutton never accepted Cavendish's result, and he became alarmed that in discussions of the subject the Schiehallion experiment, which he considered superior, was tending to be sidelined. It also annoyed him that when the Schiehallion work was discussed there was a tendency to call it Maskelyne's experiment, his own name being omitted. Writing to Cavendish didn't help, but Hutton managed to interest John Playfair, Professor of Mathematics at Edinburgh, in revisiting – literally – the Schiehallion experiment by doing a lithological survey of the site. This would result in more accurate estimates of the distribution of mass within the mountain: where was the schist, where the quartz.

In 1778 he had assumed the hill had the density of 'common stone' and found on that basis that the density of the earth was 4.5 times that of water, though he had urged a repetition of the experiment and two years later published a paper aimed at improving future experiments by finding the best place to put the observation stations, where the attraction of a wedge-shaped hill would be greatest. He had also pointed out that to get a really good result would require examining the interior of the hill by boring holes in it ('after the manner that is practised in boring holes to the coal mines from the surface of the ground'). In the 1808 *Abridgement* he then modified his 1778 paper, pushing the density of the earth up to nearly five times that of water by assuming the mountain was a little more dense than originally thought.

Meanwhile in 1801 Playfair carried out, and in 1811 he published, the lithological survey Hutton had pressed him to do. He was able to raise Hutton's original estimate of the earth's density to the range 4.56–4.87 times that of water, depending on assumptions about the rock strata below the surface, but Hutton now seems to have lost confidence in these reworkings. In an 1812 reprint of his 1778 paper he suggested that a calculation based on an assumed distribution of rocks within the hill would be 'a mere useless labour' because of how much was unknown about that distribution, and he stuck with the value of 'nearly 5' times the density of water that he had put in the *Abridgement*. In the 1815 second edition of the *Dictionary* he did the same.

This new flurry of activity around the Schiehallion work succeeded in increasing the visibility of Hutton's contribution, reviewers of the works concerned hurrying to note that the calculations he had done in 1778 had been 'greater than can easily be imagined': they were 'more laborious, and, at the same time, called for more ingenuity than has, we believe, been brought into action in any computation undertaken by a single person since the preparation of logarithmic tables'. But Hutton's sensitivity on the point was itself becoming a matter of comment, and a full recalculation based on Playfair's new survey of the site had not yet been done. There, for a while, the matter rested.

∞

Amid all this attention to his own old texts Hutton might seem to have given up producing new work. Yet he did have new scientific papers on his desk, and the old problem remained of where to publish them.

Over the years some projects were simply abandoned. His furious rate of work gives an impression of tremendous efficiency, with nothing wasted and nothing thrown away. But the reality was different, and by the time of his retirement Hutton's study was littered with the remnants of work that he had not seen through to completion or publication. A project to construct trigonometric tables in radians rather than degrees, for instance: back in 1784 Hutton had solicited helpers to work on the calculations, and he was at one stage giving some thought to how to organise the several assistants who were 'closely engaged' in the work. The project stayed warm for a few years, but it never appeared in print and by now it was clear that it never would. Likewise, Hutton had obtained the papers of the Royal Society's librarian John Robertson after his death and, from these, put together a substantial amount of information about mathematical instruments, including historical notes and lists of old works on the subject. He copied the texts out on the backs of cut-up sheets of calculations, but he never got around to publishing them. Again, he made lengthy selections and abridgements from the works of Archimedes and Pappus 'to exhibit the elegance and force of the ancient Geometry'. And again, no publication ensued.

Some pieces, though, were ready for publication and were languishing simply for want of a suitable venue, or perhaps for want of the time to liaise with a publisher and organise the text into the right size and shape. There was a book-length translation from the sixteenth-century Italian mathematician Niccolò Tartaglia, on the solution of cubic equations; made around 1790, this had been used as source material for Hutton's history of algebra in the *Dictionary*, but the translation itself had not yet seen print. There were a few new results about infinite series, including methods for finding the lengths and angles of a triangle from incomplete information, without the use of tables. And there were a few new explorations of geometry, including some on cross-sections of the spheroids, conoids and hyperboloids that had long featured in his *Mensuration* and other works.

Furthermore, Hutton had continued to pursue his historical work, amassing new material on the history of algebra: new details about British and European mathematicians, but also a mass of information concerning Indian and Arabic algebra, based on recent publications in the *Asiatic Researches* and elsewhere. He had seen specimens of Indian algebraic works both in Persian and English translations, sent by Edward Strachey of the East India Company to a mutual acquaintance; Hutton also copied out a series of relevant papers on the subject by his one-time rival Reuben Burrow.

The treatise on bridges, too, had generated new matter Hutton wanted to publish; in 1801 he had authorised a reprint of the original text, but he wished both to respond to the early critics of that text and give some new proofs. Again he had exercised his historical interest, compiling a history of iron bridges over the twenty-odd years of their existence. There was an urgency to this particular project, since the technology of iron bridges was still poorly understood; some of the specimens of which Hutton had obtained descriptions he reckoned to be in danger of falling, while a few had already fallen. Where the force went in such a rigid structure, and how it was different from a construction made of separate stones, were subjects on which Hutton felt he had valuable observations and theories to contribute. A compilation of descriptions and drawings might at least do something to help avoid engineers repeating the same mistakes.

Most obviously of all, there was the final account of the work on ballistics, unpublished (at least in English) since the last experiments in 1791. It was large enough to be a book by itself, with page after page of detailed descriptions of the experiments and lengthy derivations from them of formulae and rules. Hutton had also prepared a separate final discussion of practical problems in gunnery, considered in the light of his new results. Quite probably some of this material had circulated at Woolwich, but a wider public had no idea it existed.

As he had done in the 1780s, Hutton eventually adopted the obvious solution of publishing a collection of these papers – though by no means all – in a volume of his own. Or rather three volumes: there were so many of them. He again called the work *Tracts*, and there was perhaps a vague sense that this was a continuation of the long-abandoned project for a periodical to rival the *Philosophical Transactions*.

But it was more than that. For Hutton the new *Tracts* had something of the character of a final legacy: a compendium of all the original research he had ever done. He stated in the preface that it would probably be his last original work. As well as the new matter, chief among it the ballistics work, he reprinted all of his papers from the *Philosophical Transactions* of the 1770s and 1780s, the history of algebra from the *Dictionary*, the history of logarithms from the introduction to his *Tables*, and even a few of what he judged his more interesting pieces from the 1775 *Miscellanea Mathematica*. It was an opportunity to reinvent some of that material, and by judicious editing to form it all together into a whole. It was also an opportunity to modify some of his judgements and in at least a few cases to downgrade the credit he gave to others. Finally, Hutton the schoolmaster could not resist including in this research publication a concluding series of exercises and worked questions about the material.

It was all compiled and ready by July 1812 (some new matter on the history of algebra was added at the last minute, disturbing one tract's structure); the result was a miscellany of thirty-eight 'Tracts' that defied summary. A fat set of three books – 1,300 pages in all – that were laborious in the extreme to print and to proofread, with Greek and Arabic notation in places, as well as mathematical symbols. Hutton called in a favour from the oriental librarian at the East India Company, who supplied some pieces of Sanskrit type and an explanation of the accompanying notation.

The books were a true monument to Hutton: to his work, his achievements, and the style of mathematics he valued. Reviewers, on the whole, were kind, and there were some long and prominent notices as readers laboured to pick out which parts were the most interesting and the most significant. Playfair in the *Edinburgh Review* took the opportunity for a little adulation. For him Hutton was one of those who 'have the most contributed to the diffusion of mathematical knowledge in this island' in the last half century. 'A popular as well as a profound author', Hutton was fully deserving of his celebrity and his success.

11

Controversies Old and New

8 May 1814. Off Spithead, in the ship known as the horrible old *Leopard*. Charles Blacker Vignoles – Hutton's grandson – is writing to his fiancée, and in a reflective mood.

Till within the last year I lived with my Grandfather, a Man celebrated in the literary world: a Philosopher, a Mathematician of the first rank in the present age: one to whom the scientific world is indebted, and who by his own natural Genius and Talents raised himself to what he now is – under the eye of such a Man and associating with the literary friends who frequented his house . . .

So many brilliant flashes of natural and genuine wit; so many quaint and original Ideas upon classical and literary Subjects, as met my Ear in the Attic Meetings, which I once had the honor of attending – They were held at the House of Mr Pxxxxn of Burners Street: His Daughter Ellen a young Lady of the most surprising Genius and Talents, was the chief attraction: she was far from being handsome: a petite brunette with plain features: but a pair of fine black eyes, uncommon good humour, a flow of animal spirits and an extraordinary share of ready talk & repartee attracted all her Fathers literary friends[.] Among these was the celebrated sculptor Fxxxxxn, and his Family. This extraordinary Man is deformed, and what is very singular his Sister, (and I believe his only one) is also a sharer in that apparent misfortune, which is the Ridicule of Fools and Children.

Charles Hutton around his retirement.

F was the soul of the meetings. A naval man brought home a curious piece of wood which after discussion was formed into a small chest and called facetiously **H ΘEKE ATTIKE**, the attic chest. Contributions to the chest were sent in anonymously, chosen and read every second Tuesday from nov to 17 jul.

∞

Yes, Charles Hutton was admired, revered, loved. In retirement his round of friends and colleagues was happy, even brilliant – as Vignoles recalled – and his own household was distinguished, charming and popular.

But others weren't so sure, and during the years of his retirement he was dogged by a series of controversies and old scores. It's rare to attain a position of Hutton's eminence without making a few enemies, leaving a few disgruntled former colleagues such as Reuben Burrow along the way. In a similar incident, William Saint, a former teaching assistant at the Royal Military Academy, launched in 1810 a slashing attack on that institution and its educational standards. He wrote to William Mudge, by now the governor of the Academy – several times – and when he received no satisfactory response he printed his letters as a book. The state of studies was, for him, 'miserable and abject' and cadets were graduating ludicrously ignorant.

Evidently some of his ire arose from his own bad experiences as a teacher. He had resented both the dominance of Hutton's *Course* and a series of orders that required him to stick to it in his teaching. After making a certain amount of fuss he had received a verbal order stating that he was at liberty to instruct as he pleased; but that a knowledge of the *Course* would still be required at examinations.

While never actually naming Hutton as the author of the Academy's (alleged) troubles, he insinuated that he had presided over a total ossification of the syllabus, with the students doing little more than copy out sections of the printed *Course* and cram them without understanding. Mudge as governor was also at fault, of course, for his 'disgraceful management' of the institution under his care. The sister establishment, the Royal Military College at Great Marlow, incidentally, taught nothing but 'idleness and depravity'.

Saint was one of those people who by their vehemence tend to obscure matters. It's hard to tell whether there was any truth in his charges, and if so, where it lay. It's possible that a too exclusive reliance on Hutton's *Course* had arisen, that uniformity of teaching was being secured at the expense, at least in some cases, of students' understanding what they learned. A profusion of new orders about the teaching of mathematics at the Academy in the few years after Hutton's retirement in 1807 suggest that others than Saint thought something was amiss. And there seems some ground for his additional charge that with nearly two hundred cadets now at Woolwich, the Academy was again seriously understaffed. But that the college was rotten to the core seems unlikely, and Saint's evidence suggests instead that little had changed since the mid-eighteenth century: some of the cadets didn't understand all the material they trotted out in their examinations, some were idle (indulging in games, in 'loose and filthy conversation, in making obscene drawings upon their slates and books, in caricaturing the masters') and a few cheated.

Saint's attack may have wounded Hutton personally, but it produced no other consequences; the drastic reforms he wished for did not take place, and Hutton's *Course* remained the foundation of the syllabus at Woolwich and Marlow as well as at a number of other military schools. That the governor of the Royal Military Academy shortly afterwards asked Hutton to prepare the new third volume of the *Course* was a clear mark of his confidence in Hutton and his book. William Saint's subsequent career is obscure.

Saint's opposition to Charles Hutton, like Burrow's before it, had something of farce about it, but from a different direction there came a far more serious and damaging threat. As early as 1802 it had become apparent that the mathematical reviewer for the widely read *Monthly Review* was waging a one-man war against Charles Hutton and the kind of mathematics he stood for. A distinctly mixed review of the *Dictionary* ('not remarkable for the accuracy of his definitions') had appeared in 1798, and in March 1802 came a frankly rude one-page notice of the new edition of Hutton's *Principles of Bridges*. The reviewer alluded to the 'unsubstantial systems which the pride of calculation is continually erecting', to the 'triflings' of 'speculative men'. He implicitly accused Hutton of plagiarising fellow northerner William Emerson, and he stated that the book did not answer the purpose for which it had been (re)published. The piece was anonymous, as nearly all reviews were, but it was no secret that the main, virtually the only reviewer of mathematical books for the *Monthly* was Robert Woodhouse: a Fellow of Caius College, Cambridge and a most able mathematician (he had been senior wrangler – highest-placed in the examinations – in 1795).

Woodhouse's insults were strong stuff, and Hutton wrote in some wrath to the editor of the *Monthly*, asking that his (enclosed) rebuttal be printed. The editor consulted Woodhouse and decided to stand by him, declining to print Hutton's piece. Woodhouse seems to have pointed out that some of his criticism did not make it unambiguously clear whether he meant to attack 'speculative men' in general or Hutton in particular, and that he had not *explicitly* accused Hutton of plagiarism. Flimsy, as Hutton pointed out: no reader of the offending article could possibly have doubted what Woodhouse meant. Hutton was sensitive at this stage in his career to the issue of damage to his reputation ('some consideration is due from you in such a case for the feelings of an author so outraged') and the two closely, rapidly written sides of his reply breathed indignation. Finally he asked for the return of his rebuttal so that he could print it elsewhere; it eventually appeared in *The Monthly Magazine* as a short essay on the 'abuse of reviews', while in a sort of compromise Hutton was allowed to insert a shorter piece in *The Monthly Review*, to which Woodhouse again replied. Hutton also asked the editor not to have his future books noticed in *The Monthly Review*, but this was ignored, and Woodhouse was allowed to write lukewarm reviews of the *Recreations* in 1804 and of the *Abridgement* ('in some places we discern something like caprice') in 1805.

On one level, all this looked like another merely personal attack. But, as Hutton said, there was no evident motivation for it, and he speculated that it was about the snobbery of Cambridge mathematicians ('a general propensity . . . to affect to despise, & endeavour to disparage, whatever comes not from their own society'). In fact, he most likely knew perfectly well that it was about deeper issues: issues which would become increasingly visible over the next decade, and increasingly worrying to Charles Hutton.

What it was about was the right way of doing mathematics, and the value of the distinctive British methods that had developed over the last fifty years or more. That was the mathematical culture to which Hutton had given his life: steady, amateurish, proceeding in the pages of *The Ladies' Diary* and *The Gentleman's Diary* to make small incremental additions to mathematical knowledge, in a framework laid down in large part by Isaac Newton back in the seventeenth century: celebratory of Euclidean geometry, which was a key part of British grammar school and university education, and reliant wherever possible on geometrical intuition, particularly in handling subjects like mechanics or astronomy.

'A certain degree of mathematical science, and indeed no inconsiderable degree, is perhaps more widely diffused in England than in any other country in the world,' wrote John Playfair in 1808, pointing to *The Ladies' Diary* and the other philomath periodicals as evidence. As Hutton himself had put it in 1775, 'By means of such problems, and little essays, considerable additions are made to the stock of mathematical learning in general, as well as to the particular knowledge of individuals.' Such additions were not to be despised, and by its richness in them British mathematics was successful in its own terms. And it produced material such as Hutton's ballistics and his *Course* that were successful as exports, taken seriously as research material or as pedagogy in Europe, America, and the British colonies in India. It produced mathematicians perfectly capable of working and teaching both at home and overseas, and it supported British engineers and engineering through a period of unprecedented change and innovation in the later eighteenth century. Moreover it produced men like Charles Hutton himself, taken seriously as author, colleague and correspondent by mathematicians across the world.

France, and the European continent generally, had gone a different way. There, the learned academies supported a small number of elite mathematicians to carry out intensive research. They had no real counterpart in Britain: certainly not in the relaxed figures who held mathematical teaching posts or professorships in the British universities. Since the 1740s these well-funded, state-employed mathematicians at Paris, St Petersburg and elsewhere had been pursuing a special line of thought. Called analysis or infinitesimal analysis, its effect was to remake the calculus without the aid of geometrical or dynamical intuition. From the outside it initially looked like a highly abstract game with little point to it: playing on the edge of philosophical impossibility with the infinitely small and how it should rightly be studied. The sort of thing Fellows of the Paris Academy or favourites of the Tsar were free to indulge, but of little real interest to the working, practical mathematician.

But by the end of the century the outcome was that on the European continent the study of force and motion – the phenomena that the calculus was ultimately about – had been placed on quite a new footing. And mathematicians there were increasingly doing things the British could not match in applied fields such as astronomy. Britons like Hutton kept an eye on all this; he had continental mathematics books in his personal library by the yard, and he subscribed to all the main continental scientific journals: as did others, and as did institutions like the Royal Society. His *Dictionary* was full of admiration for, and full of references to, new continental work; so were his *Tracts*. Other Britons followed him in this; some published translations of continental works, others tried actually using the new continental notation and conceptual language in their own writing. Hutton felt unable to go that far; perhaps constrained by his role and background as a teacher, and unwilling to create unnecessary difficulties for his readers.

For a vocal minority among the coming generation of British mathematicians – and Robert Woodhouse was one of them – all this wasn't enough. They cultivated a sense of crisis, of outrage, of national shame at what France was doing and Britain wasn't. They wanted change, fast: thoroughgoing reform of research, publishing and teaching. When Laplace published his *Traité de mécanique céleste* in 1799, showing that France was now leading the way in the very areas of mathematical astronomy where the great Briton Newton had done so much, they reacted with something close to panic.

During the last half century, [in] the mathematical sciences . . . scarcely any improvement has been made in them in England.

A mathematical production, above the level of school-practitioners, finds little encouragement in this country; to enable a book to sell, it must be trifling; it must reduce all rules to mere mechanical operations; it must in fact be suited to the taste of solvers of problems, and not to investigators: – we have more of the former class, and fewer of the latter, than any empire in Europe.

Even Hutton's friend and admirer John Playfair, in a much-quoted review of Laplace, joined in the criticism of British mathematics. Despite the many good things he had to say about the thriving state of British mathematical culture, Playfair pointed out that there were only a handful of Englishmen who could read recent French mathematics with understanding, and he saw that as self-evidently a bad thing:

> A man may be perfectly acquainted with every thing on mathematical learning that has been written in this country, and may yet find himself stopped at the first page of the works of Euler or D'Alembert . . . from want of knowing the principles and the methods which they take for granted . . . If we come to works of still greater difficulty, such as the Mécanique Céleste, we will venture to say, that the number of those in this island, who can read that work with any tolerable facility is small indeed.

Over the next few years some of the critics of British mathematics were to form the short-lived Analytical Society at Cambridge, to press for the reform of mathematical teaching there. They published a translation of Lacroix's French calculus textbook and other material, and promoted the story that British mathematics was in crisis.

On one level it was clear that much of their rhetoric was exaggerated and self-serving; and the sudden reforms they called for didn't happen, were in no particular danger of happening. But at another level it was clear they had a point: that British mathematics was now being transformed – albeit slowly – by its contact with continental mathematics. That process acquired new momentum after the peace of 1815 made communication across the Channel a great deal easier. New research agendas arrived to stay; continental mathematics appeared in more and more English translations and summaries, and received more and more British responses, for instance in the pages of the *Philosophical Transactions*. And continental notation and concepts became increasingly visible even in native British mathematical publications.

These were changes in which Hutton, in his seventies, was not likely to participate. He had done more than anyone living to define, shape and nurture the British mathematical culture that was now under attack. In the short term Hutton could try to limit the damage. He could write letters of protest to editors when his works were criticised, he could publish rebuttals and try to avoid sounding merely peevish when he did so. He could brief Olinthus Gregory to attack in his name – though Gregory himself was not wholly unsympathetic to continental-style mathematics. But Hutton could not, either by private remonstrance or public rebuttal, alter the fact that mathematics was changing, and that it now seemed he would live to see much of what he had achieved become out of date, even irrelevant.

Of controversy, before and during his retirement, Hutton thus had more than enough from his colleagues. Unfortunately his family managed to provide a large dose too. Charles Blacker Vignoles, the young man plucked from Guadeloupe and welcomed into Hutton's household in 1795, was now approaching his twenties. Hutton had provided for him, had overseen his education. There was a period at a school in Kent; there was probably private tuition from Hutton himself. Vignoles acquired a grounding in mathematics, classics and modern languages. For a long while he was a lively and much-liked member of the household. He showed talents for music and drawing; he wrote verse, plays and music; he sang and recited. His (step-)grandmother and his aunt Isabella doted on him, and in 1806 a cousin remarked that he seemed such a boy as might be expected to distinguish himself.

But things were not altogether so happy. Vignoles perhaps tended to be somewhat too carefree and, as he grew a little older, 'too fond of a dash, a show-off', as he himself put it. He had the run of his grandfather's library, but used it for pleasure rather than instruction and Hutton, pleased to see the boy reading at all, apparently neglected to regulate *what* he read. As an adult, Vignoles would recall working through logarithmic and other tables for Hutton's published works; this must mean checking proof-sheets or perhaps doing recalculations and enlargements of matter for the later editions of Hutton's *Tables*, or the tables in his *Tracts* or *Course*. Typical of the work members of Hutton's household had been doing since the 1780s, but hardly the most stimulating activity for the young man.

Hutton, for his part, was not entirely able to forget the expense or the disruption young Vignoles had brought with him. Several manuscripts show Hutton attempting to document the sums Vignoles had cost him, even to recover some of them from the government through a regulation providing recompense for items lost when officers died on active service. He wrote around distant acquaintances in pursuit of rumours about grants of land made to the Vignoles family in the Isle of Man, France, North America.

Reading between the lines of Vignoles's letters, Hutton also seems to have demanded too much of the boy, to have pinned on him too many of his hopes. Hutton's own son had had a distinguished career in the Royal Artillery but had not made a splash intellectually. Two daughters were dead; one – Eleanor – was married, and we hear little or nothing of her and her children in Hutton's letters, while faithful Isabella was Hutton's devoted amanuensis and companion but – perhaps in part for that very reason – seems to have developed few independent intellectual interests. Charles Blacker Vignoles represented Hutton's natural hope for a certain kind of intellectual legacy.

The law, it was decided, would be his career; whether he had any say in the matter is not clear. He was placed under articles to a proctor in Doctor's Commons: a sort of legal apprenticeship. With peace looming, this did indeed promise a more secure future than the perhaps more natural choice of the Army, where Vignoles had a family tradition and Hutton many personal connections. And the Inns of Court, home to generations of city lawyers, were just around the corner from Bedford Row. But Vignoles did not thrive. By the age of twenty he appears to have been quite desperate for escape of every kind. The situation had become an explosive one, and in the summer of 1813 it duly exploded.

There was, Vignoles later hinted, an incident with a girl: or rather two girls. The details (naturally) are lost, but it seems that what he himself called his 'extravagance and imprudence' coupled with these 'boyish inclinations' to elicit a 'thunderbolt' from Charles Hutton. Vignoles, for his part, insisted on abandoning the law, breaking his articles and starting again in a military career.

The outcome was that Vignoles left Bedford Row in about June 1813. He went to lodge with Hutton's colleague Thomas Leybourn, mathematics master at the new Royal Military College at Sandhurst. He wasn't a cadet; strictly speaking he still held a half-pay commission in his father's regiment, awarded to him in infancy as a sort of compensation for his parents' deaths. The intention was that through private tuition from Leybourn and some contact with the cadets Vignoles would acquire enough learning and demonstrate enough willingness to persuade some officer to admit him to serve in his regiment. The Leybourns were kind, the Huttons continued to provide financial support, and Vignoles had achieved the escape he seems to have craved. Hutton wasn't altogether happy about the situation; in his seventies, he had become somewhat querulous about money, and he resented the sums he had already spent on Vignoles and the new waste involved in leaving his legal training.

There followed a not unreasonable series of military adventures for Charles Blacker Vignoles. Commissioned in 1814, he was taken prisoner at the disastrous British attack on Bergen-op-Zoom and returned to England on parole. Having been exchanged and ordered to Quebec, he was shipwrecked (in the horrible old *Leopard*) at the mouth of the St Lawrence river. He was with the army of occupation in Paris after Waterloo; after repeated requests to be allowed to use his mathematical and drawing skills as a staff engineer – all refused – he was eventually made aide-de-camp to General Sir Thomas Brisbane at Valenciennes.

The family watched with some consternation. Margaret was desperately worried for Vignoles, whom she thought of unhesitatingly as her grandson, and a cousin wrote with tears in his eyes of his hope of seeing him again. But as well as the obvious dangers there was on Vignoles's part an equally obvious enjoyment of the new life. He left a manuscript tragedy behind him at Sandhurst, and an unfinished comedy; he wrote an eight-canto *Sylphiad* indebted to Alexander Pope and made quite a hit at regimental theatricals in Quebec, starring in no fewer than four female parts including Julia in *The Rivals*. Fond of 'balls, routs, concerts, French and English plays', he calls irresistibly to mind some of the more unsteady young men in Jane Austen (Henry Crawford, perhaps?).

But. But. Like a thousand other lieutenants he was put on half pay in the reductions after Waterloo. The blow struck him personally in January 1816. His military career had lasted just two years, and he now despaired of its future. He had been a secretary to the Duke of Kent for a time, and both he and the commander in chief (Frederick, Duke of York) made vague promises to him. But they had many claims on their interest, and there simply wasn't much work to go around compared with the glory days of sweeping Boney out of Spain and France. Vignoles chafed, he raged, he regretted, but there was little he, or anyone, could do.

∞

All might yet have ended in the reconciliation that most of the family wished for, but the years 1815–16 saw not comedy for Hutton but both family tragedy and further professional reversals.

Henry's son Charles Henry, born in 1800, moved to Woolwich in early 1814 to study at the military academy where his grandfather had taught for so long. Conceivably the example of cousin Charles Vignoles contributed something to the decision to pursue a military career at an unpromising moment. Briefly he appeared to be thriving there. In August he wrote home proudly of his passage from one class to the next and his hopes of soon being a corporal. He was studying under John Bonnycastle and Olinthus Gregory, old friends of his father and men who surely showed him some favour on that account.

But then it all went wrong. From one month to the next he sickened of the life at the Royal Military Academy, took 'quite a hatred to Woolwich, and the life of a soldier'. He sickened physically too. Consumption is spoken of in some documents, but precisely what the matter was is nowhere recorded. A bad cough, a loss of weight.

In October he was moved to Bedford Row, his life despaired of. Hutton spared no expense on his comfort or his treatment. He called in two expensive doctors; they said that nothing could save the boy, but he nevertheless insisted they attend every day until the end.

Charles Henry Hutton lingered for five months. He died on 13 February 1815. He was buried in the plot at Charlton where lay Charlotte, the aunt he had never known.

Demoralised perhaps by the attacks from Saint, Woodhouse and others and exhausted by the vicissitudes of his family life, Hutton conceived the idea that his creative scientific work was over.

Now seventy-seven, he planned to retire to a smaller house, perhaps in the country, for the remainder of his days. As a necessary corollary he set about disposing of his library. 'I shall have little or no further use for it, and it would prevent me from chusing another residence, and . . . it could be of no use to any of my family, after my death.' (A rather devastating judgement, that, on the intellectual attainments of his children and grand-children.)

He had been accumulating books since the 1750s, and the mathematical section of his library was now generally acknowl-edged to be the best in Britain; it contained over three thousand volumes and included very rare early printed works on algebra and geometry, unique collections of editions of Euclid's *Elements of Geometry* and of Newton's *Principia Mathematica* among other treasures, and almost certainly the best private collection of conti-nental mathematics books in the United Kingdom: more than 130 volumes from Clairaut, Euler, Lalande, Laplace and many more. He had collected manuscripts from friends and from the sale of their collections; had annotated some of his books and pasted letters from friends and colleagues into others.

Hutton's idea was to sell the collection complete to the British Museum, which had a fine library of its own. There were prece-dents; during his time on Council at the Royal Society, Hutton had seen the Society's collection of artefacts transferred to the British Museum in an amicable and – at least in intention – a mutually beneficial arrangement.

Letters came and went; it seemed the Museum had few mathe-matics books and the trustees would be happy to augment them by taking Hutton's library. It was agreed to abide by a valuation deter-mined by one representative from each party; the Museum appointed an officer to inventory the books, who reported favourably.

But one of the trustees of the library, presently away from the capital in Lincolnshire, was Sir Joseph Banks. Fearing he might take offence if not informed of what was going forward, Hutton wrote to Banks, expressing the hope that he would approve of the proposal. No reply. But within a fortnight Banks was back in London, and the deal was off.

Hutton and his friends took this as proof, if any were needed, that Banks and his circle still held a grudge against Hutton and his mathematical friends, that his old enmity was, as Hutton put it, implacable. Hutton felt his treatment was cruel, and he was angered, perhaps disproportionately, by what he imagined as Banks's triumph over him. Instead of casting about for another institution that might take his books (there were several) he determined to sell them off at auction.

Isabella first worked through the library making a draft catalogue, to which Hutton then made additions and changes. He made a note of the books he would like to keep: something like one in ten. In November 1815 Hutton himself made a fair copy of the catalogue – changing his mind in complicated ways about what to keep – which was turned over to the firm of Sotheby's for printing. The printed catalogue, which filled eighty pages, was available all over the country, and the sale was well attended. It took six days to sell everything, in the middle of June 1816.

The dispersal of the library was greeted with horror when it became known to Hutton's mathematical friends. It was, indeed, a real blow to the British mathematical community, which might otherwise have gained an excellent publicly accessible library of its own. The collection was twice the size of Playfair's and nearly three times that of Maskelyne. The sale was surely a melancholy event to Hutton too, and it's hard to understand quite why he persisted with it. For he didn't, in the end, retire to the country, nor did he cease his work, continuing to produce new articles and new editions of his books for most of the next decade. Why then sell the books? Defiance? Despair?

But still, the books went. Friends of Hutton acquired at least a few, and it was possible at the last minute to make sure some went to the Literary and Philosophical Society of Newcastle. A London bookseller named Weale also bought a few, bringing them together with items from the collections of Horsley, Maskelyne and others which he later resold. But most simply dripped away lot by lot, who knew where: 170 bound volumes of almanacs, going back to the early seventeenth century; sixty-odd volumes of Hutton's own works, and over 150 offprints and pamphlets, some with his own manuscript corrections. Many of his personal friends, favourites and former favourites, were represented, some by fancy presentation copies and some by letters bound into the books: Maskelyne, Bonnycastle, Emerson, Ferguson, Franklin, Gregory, Leslie, Martin, Muller, Priestley, Simpson; a roll-call of the British scientific culture Hutton had known. Perhaps Hutton's sense of the transience of that culture was part of the point.

Manuscripts went too: a medieval Bible, works on mechanics, gunnery, astronomy. Most intriguingly of all, sold with the books was a collection of the mathematical instruments of Benjamin Franklin. Had Hutton known the great American scientist and statesman, in a relationship otherwise lost from the historical record? Both were close to John Pringle at the Royal Society in the 1770s, but of other evidence we have none.

Some frivolous works were sold, as well. *Cynthia: with the Tragical Account of the Unfortunate Loves of Almerin and Desdemona. Every Man His Own Brewer*. So were what seems to have been the very last of Hutton's childhood collection of border ballads: Allan Ramsay's *Poems* and *Scots Poems of Before 1600*; the *Gentle Shepherd*, *Caledonian Miscellany* and *Tim Bobbin's Toy-Shop Opened*. But on the whole, non-mathematical items in the sale catalogue were conspicuous by their rarity, and Hutton apparently hung on to much of what he, Catherine or Isabella might find entertaining. Absolutely none of their collection of music was sold.

By the spring of 1817 Hutton was lamenting the sale, wishing he had kept the books together in a sale or gift to another institution. He went on lamenting it in his letters for five more years, but it was too late.

∞

Meanwhile, the mourning for young Charles was scarcely over when Margaret Hutton, too, sickened. She was in her early sixties, and already in the summer of 1814 she had been 'very unwell', telling Vignoles 'I never expect to injoy sound health more'. She was correct, and although the winter of 1816 saw her still avidly reading and corresponding with Vignoles and others, by March 1817 she was dead.

Margaret's death is as frustrating for the biographer as her life. Vast though her role in Charles Hutton's life and his work undoubtedly was, and highly visible though she certainly was to him and everyone who knew him, the documentary sources have almost nothing to say about her. A handful of letters; a poem (the one about Samuel Horsley at the Royal Society); the inscription on her tombstone (now destroyed) and the licence for her marriage (its details partly false). Handwriting that is probably hers in some of Hutton's mathematical and scientific manuscripts, assisting, revising, correcting. We don't know what she died of or how long her final illness lasted; we don't know for certain when her funeral took place or who attended it. Even her exact age is unknown, since the dates given in her marriage licence and on her tombstone are inconsistent. Like so many Georgian women, she is only partly visible in the sources we have.

Be that as it may, Hutton was devastated. Margaret had been the mainstay of his life since shortly after his arrival in London forty-four years before, and they had weathered together the catastrophe of Hutton's career in 1784, the death of their only child, the unexpected arrival of young Vignoles and his stormy departure. She had assisted with his calculations and his prose, and we shall never know how much of his success and sheer ability to get work done he owed to her. Proud of her husband to the end, she wrote to Vignoles in 1814, 'Let it not be said my dear Charles, that a grandson of Dr. Huttons lived in vain. a meer ordinary being.'

Her death left him, as he put it, 'almost alone' in the world. The bustling household of a few years before was now reduced to himself and Isabella, with a cook and a maid. In August 1817 Charles Hutton celebrated his eightieth birthday – if he celebrated it at all – in a bitter key.

∞

It was perhaps fortunate that Margaret had never known the whole story of Charles Vignoles, for a few months later a new thunderbolt arrived from that quarter. An attractive ward had been staying with Thomas Leybourn at the time of Vignoles's arrival in the summer of 1813. She was Mary Griffiths, the eldest daughter of a Welsh farmer; she had worked as a milliner in London and cared for her two younger sisters after their parents' death. Twenty-six to Vignoles's twenty at the time of their meeting, she shared his emotional and romantic temperament. By the end of September that year they were secretly engaged.

There ensued a voluminous clandestine correspondence, with all the ingredients of quarrel and reconciliation, misunderstanding and forgiveness. The sometimes feverish tone wasn't helped by the infrequency of their meetings. Mary had no wish to enter upon a penniless marriage, but Vignoles had conceived a plan to go to Central America with the revolutionaries (you could hardly make it up) and would not leave her behind without making their union permanent. So clandestine correspondence was capped in July 1817 by a clandestine marriage; Mary posted down to Portsmouth overnight, and early on Sunday morning, the thirteenth, they were married in the village of Alverstoke. A fortnight later Vignoles left for America in the desperate hope of making his fortune, leaving his bride to inform her family and his.

When the news broke, both families were appalled, and Hutton – who had paid for Vignoles's American equipment and had been up to this point at least sporadically 'amused' by the young man's letters and news – refused to have anything more to do with his grandson, refused even to have Isabella read aloud his letters to her. No more money, no more favours: nothing.

Hutton cannot altogether be blamed for taking a dim view. Vignoles was twenty-four; he had entered upon a secret marriage with a woman he had no means of supporting, and quit the country leaving her pregnant in cheap lodgings. His American plans were little more than a fantasy – they changed several times and he eventually went not to join the rebels but to work as a surveyor in South Carolina and Florida – and his ability to support even himself appeared utterly remote. From Hutton's point of view, if the young man had been trying to look like a wastrel he could scarcely have done more, and it made matters worse when an offer of army employment came in just weeks after his departure, highlighting the fact that his own hotheadedness was largely to blame for his situation.

But Hutton does seem to have overplayed his role of outraged older relative. After the initial burst of rage, Leybourn wrote a few months later to offer Vignoles his forgiveness as Mary's former guardian. Hutton did no such thing, remaining steadily implacable as months lengthened into years, and refusing to do anything for Mary even when she was in real want; the Leybourns took her in for a while when she was ill after the birth of her first child.

The affair became an open wound in the Hutton family. His aunt Isabella sympathised with Vignoles; so did aunt Ellen. Isabella wrote covertly, sometimes from the address of a friend, and enclosed a five-pound note when she could. Henry Hutton sided with his father, and would have nothing to do with a young man he thought culpably ungrateful as well as irresponsible.

Should Hutton have acted otherwise? He too had once abandoned a family-chosen career in favour of a more independent path, and he too had married a woman of whom – probably – his family disapproved. He too had used physical relocation to indulge in self-reinvention, leaving wife and, for a while, children behind in the process. Yet he had prized financial security and provided handsomely for his dependents; he had moved *to* jobs, not away from them. He had no reason to admire or to condone Vignoles's conduct, and every reason to think it would end in disaster. He could have helped more than he did, but it would surely have required a superhuman magnanimity to take on the support of Mary, and perhaps an uncharacteristic recklessness to do so at a time when he realised he himself had not very much longer to live.

12

Peace

Bedford Row, London. 21 September 1822. A small group clusters around the door of one of the handsome tall houses. A knock; the maid goes to fetch the master of the house; and white-haired Charles Hutton appears. A little slower now; a little deaf. His piercing eyes take in his friends. Olinthus Gregory, whom we know; Francis Baily, astronomer; Dr Andrew, one of the teachers from Addiscombe; a few more.

Once inside, one of them reads from a paper. We have the honour, Sir, of waiting upon you. Respect . . . admiration . . . veneration . . . gratitude. He unveils a marble bust of Charles Hutton, the fruit of subscriptions from over a hundred well-wishers, and 'a testimony of respect for your virtues and talents, and a tribute of gratitude for your important labours'.

There's no surprise to the scene, which has been long in the planning. But still, Hutton is moved almost to tears. 'Nothing could be more gratifying to my feelings than this demonstration of your regard.' 'If . . . any thing could enhance the value of this Gift, it is the kind manner in which it is now presented. It is not in the power of any language to express my gratitude.'

∞

The decade around Waterloo had thrown Hutton an almost unbelievable series of new misfortunes through his continuing anxiety about his reputation, his intellectual legacy and of course his family. But there followed a period of something very like peace.

We hear nothing more about Hutton's plan to retire to the country; he stayed in Bedford Row. In 1821 the (now retired) General Henry Hutton moved to London to be near his father, bringing his second wife and second son. Ellen was occasionally around; widowed, she had married a Captain Wills in 1814 and visited her father from time to time. Hutton was reportedly 'delighted' by the society of his children, and affectionate towards them, particularly to devoted Isabella who was now his constant amanuensis for letters and other writing. His own hand he judged a 'tottering scrawl', though on occasion he did attempt a short letter himself.

It was still a bookish house. 'Cousin' Catherine kept Charles and Isabella supplied with copies of her own novels and other writings, and she recommended other reading matter too, including the highly regarded novels of a Miss Austen. A portrait of Hutton in his later years gives a strong impression of cosy domesticity, complete with cap, book and fireplace.

There were plenty of friends visiting and to visit, and the Huttons had still quite a social circle. In Bedford Row itself, social evenings involved poetry and music, lectures and discussions. There was an unending stream of scientific or semi-scientific visitors to entertain and be entertained by: the kind for whom Hutton would make notes in advance about topics of conversation – Olinthus Gregory, Thomas Leybourn. If there were fewer plays, skits and musical compositions since Vignoles had left, there was still conversation and companionship, and wit in plenty; and failing that, a quiet game of cards.

Hutton was increasingly cheerful in his old age; he seems, even, to have had a reputation for a sense of humour. One day an unknown wag summoned a huge number of tradesmen to attend him: horses, chaises, coal dealers, physicians, accoucheurs, apothecaries, every article of luxury or utility that could be thought of. The coal waggons alone nearly blocked the street, and the prank went on all day as different trades came and, disappointed, went. There's no record of his reaction, but evidently someone thought Hutton would see the funny side.

It wasn't all fun and games; there was always more work to do, and Hutton went on helping to promote mathematics and learning in ways he thought were valuable. Working, through his personal connections and the continual revision and re-presentation of his writings, to disseminate the values he thought important in a changing world. Friends sent him their books and papers for comment: Francis Baily on the fixing of an astronomical instrument; a Newcastle teacher on Johann Heinrich Pestalozzi's novel system of education. Others continued to consult him as an expert on bridges. The Corporation of London wrote in 1819 about a proposed replacement for London Bridge, and he gave a judgement about such matters as the flow of the tide after the enlargement of arches or the removal of the existing bridge, the depth of water and the function of the present bridge as a sort of dam; the need for further embankments.

And a chief employment, still, was new editions of his works. The *Recreations* reappeared in 1814, the *Dictionary* in 1815: the latter with substantial additions and improvements which Hutton said had cost him 'immense labour'. There were further editions of the *Course* in 1813 and 1819–20. Some of his helpers were now gone, and it isn't clear whose assistance he had in seeing all those words and numbers through the press. Quite possibly Olinthus Gregory's, although Gregory was involved in massive projects and teaching commitments of his own. Writing in the new edition of the *Dictionary* Hutton mentioned rather charmingly that his old friend Nevil Maskelyne, who had died in 1811, 'did not publish much'. Maskelyne had published eight books, forty-nine editions of the *Nautical Almanac* and thirty papers in the *Philosophical Transactions*.

Hutton still busied himself, too, with recommendations for mathematical jobs, and with philanthropy more generally. Friends and the children of friends benefited from letters written on their behalf, and one day Gregory, on one of his regular visits, found Hutton reading a letter from the wife of a penniless country schoolmaster, 'the tears trickling down his cheeks'. – 'What do you mean to do?'

'I mean,' replied the Doctor, smiling, 'to demand a guinea from you, and the same sum from every friend who calls upon me to-day; then to make up the amount twenty guineas, and send it off by this night's post.'

At the same time, Hutton maintained his own personal networks. Friends still visited, and so did more or less distant relations. There were cousins on the Vignoles side with whom he kept in touch. Thus a Gilbert Austin – cousin of the dead Charles Henry Vignoles – and his wife and nephew were house guests in 1805, and the following year Austin sent Hutton a copy of his book on rhetorical delivery, for what it was worth. He felt 'profound respect' towards his distinguished relative. Of Hutton's own brothers, on the other hand, there is practically no sign in the documents; a Robert Hutton, who may have been one, was buried at Long Benton in 1769.

Hutton had also, over the years, kept in touch with a number of friends from the intellectual circle in Newcastle; he sent them copies of his papers and advertisements for his books, and from time to time he acquired books published in the city, including the odd work on mining. Northerners always seem to me to be slightly over-represented in the pages of *The Ladies' Diary* during Hutton's time as editor; that may have been another way to make new connections and maintain old ones.

There was a particular connection with the Bruce brothers, John and Edward, who kept a school in Newcastle and with whom Hutton corresponded. Their school in fact was on Percy Street, very close to where Hutton was born, and they had become the town's leading science and mathematics teachers. They had close ties with the Literary and Philosophical Society founded at Newcastle in 1793, as did another prominent local teacher. William Turner, the co-founder of the society and one of its most prominent lecturers, was also the Unitarian minister at Hanover Square Chapel, where Hutton and his family had once worshipped. Hutton received news of the 'Lit and Phil' with interest, asking John Bruce to send him any pamphlets that were printed from its lectures.

As well as gifts to the Lit and Phil itself, Hutton agreed to support William Turner's philosophical lectures there, the Newcastle Jubilee school and the Schoolmasters' Association, all with regular subscriptions. When the Lit and Phil decided to build new premises in 1822 Hutton was particularly interested; the site they settled on faced his old schoolhouse across Westgate Street, and he asked to see the plans as well as contributing to the building fund. The foundation stone was laid in September 1822, and one of the toasts at the celebratory dinner was to Dr Charles Hutton, 'a native of the town, who reflected the highest honour upon it by his eminence in science, by his many acts of benevolence, which shewed at once his remembrance of his native place, and his zeal for the promotion of knowledge'.

All this put in his mind the possibility of a visit to the scenes of his youth, which Hutton thought would give him much pleasure.

He first toyed with the idea around 1815–16, but events within his family quashed the plan; he returned to it in 1822 with the idea of seeing the new premises of the Lit and Phil. At the suggestion of another northern friend, the local politician William Armstrong, he planned to make the trip that summer, taking the steam boat that now connected London and Tynemouth; he still remembered with nothing like pleasure the carriage journeys of younger days.

His London friends were most alarmed, and persuaded themselves, perhaps rightly, that the exertion would be mortal for the octogenarian. They persuaded Hutton, too, and he, sadly, abandoned his plan to look once more on the land where he had grown up and learned his trade.

∞

There was necessarily something elegiac about some of these activities, and it's clear that Hutton, increasingly retired and at times at least feeling increasingly infirm, believed that the more public acts of his life were over for good.

But into his life now fell an unexpected change; one old battle was at last played to its unlikely conclusion. Dr Charles Hutton and Sir Joseph Banks were now old men. Each had his supporters; each side was capable of landing some shrewd blows on the other, in public and in private. Banks's empire had grown, and as well as his career as a successful colonial administrator he stood at the head of a learned world that now took in the Royal Observatory, the Society of Antiquaries, the Linnean Society, the Horticultural Society, the British Museum and the Royal Institution. He sat on boards, he exerted influence, he dispensed patronage. At the Royal Society his focus had been on keeping the peace, unwilling as he was to see another outbreak of opposition to his rule, but his hand was a firm one and his agenda in no doubt.

Sir Joseph Banks.

On the other hand, around Hutton and his friends a generation of mathematicians had reached intellectual maturity nourished on tales about the Dissensions of 1784 and the persecution of mathematicians ever since; for them it was axiomatic that Banks and his empire represented unreasoning prejudice against them, their profession(s), their subject and their intellectual seriousness. The group as a whole nursed every grievance and every slight, real or imagined. Olinthus Gregory, in particular, spent effort that might have been better devoted to other matters collecting folklorish examples of wrongs done to his community by Banks and his minions, instances of the President's 'petty but inextinguishable malignity'. The astronomer William Herschel, for one, thought Gregory rather too coarse in his strictures and behaviour towards Banks and his party. Banks had vexed and opposed Maskelyne at the Board of Longitude and the Greenwich Observatory, and on his death in 1811 had organised his replacement as Astronomer Royal by the more pliable John Pond as well as the dispersal of his library (Maskelyne had offered it to government for the use of his successor, but Banks talked them out of accepting). In 1812 Banks had organised criticism of the Trigonometrical Survey of Great Britain (led by practitioners recommmended by Hutton) and the eventual exclusion of reports on the survey from the *Philosophical Transactions*. In 1818 Banks remodelled the Board of Longitude with the aim of diluting opposition to him from the mathematical professors of Oxford and Cambridge. In 1820 he opposed the formation of an Astronomical Society – led by a group of mathematicians and mathematical practitioners, and thought by some to be a deliberate manifestation of resistance to the Royal Society – and pressured some of its early members and supporters to withdraw. And so on, and so on. The Society for the Improvement of Naval Architecture, the Royal Institution, the Geological Society: all, it was said, had suffered from Banks's machinations.

The passage of time has placed some of these accusations beyond proof or disproof, and it seems certain that Gregory and his friends had, on occasion, a tendency to exaggerate. Hutton himself had indulged in some merely scurrilous criticism of the Royal Society and its programme in his *Dictionary* ('torturing the unseen and unknown phlogistic particles . . . hunting after cockle-shells, caterpillars, and butterflies'). Yet they may have had a point. The kinds of intellectual merit and usefulness cherished by the mathematicians of London and England were conspicuously not enough to secure election to the Royal Society or publication in the *Philosophical Transactions* under Joseph Banks. Some tried and were excluded – mathematician Patrick Kelly and journalist William Nicholson were both blackballed – others never even made the attempt: stockbroker and astronomer Francis Baily wrote to Gregory that election would be 'but an ambiguous honour' and did not seek it.

Meanwhile Olinthus Gregory, John Bonnycastle, Charles Wildbore, Samuel Vince, William Mudge and Francis Baily all saw their papers rejected from the *Transactions*.

All this had the happy, unintended side effect of creating a sense of shared identity among British mathematicians that might otherwise never have existed, doing indirectly what the 'secession' had failed to do and creating a stable group with a definite sense of belonging. In particular, although Banks evidently found Cambridge-based mathematicians like Edward Waring easier to tolerate than members of the London circle of practitioners and educators, bridges began to be built between those two worlds. Leybourn, of the Royal Military College, joined with the Cambridge reformers in his enthusiasm for French mathematical methods; Cambridge man Charles Babbage held testimonials from both Charles Hutton and James Ivory of the Royal Military College. Hutton corresponded, during the composition of the *Dictionary*, with Waring himself.

By the late 1810s, then, it seemed that antagonism between Sir Joseph Banks's learned world and that of the mathematicians was simply a settled feature of British intellectual life. But in June 1820, Banks died. Hutton, by six years the older man, was unexpectedly the last left standing, indeed one of the very last of the main players from 1784 still alive, and to many there seemed an opportunity to put right, in one way or another, the relationship between mathematics and science, between mathematicians and the Royal Society.

Responses were swift to come. One journalist wrote, under the cover of an obituary of Banks, a swingeing attack on the mathematical party that had opposed him in 1784 and its intellectual heirs. The Dissensions themselves were now more a matter of history than of memory – and rather garbled history at that – but there could be no mistaking the target of such paragraphs as these:

The whole attainment of the mass of mathematicians consists in trivialities long discovered, useless, or beyond their skill to use, and totally inferior in the required mental vigour, in public service, and in improvement of the understanding to every other intellectual acquisition; – infinite degrees below the genius essential to oratory, poetry, or painting.

. . . Of a thousand mathematicians, not the human cube root has ever been, or will be, more than the depository of the dusty problems, that the bookmakers of the art, the Simpsons, and Huttons, and Bonnycastles, have transmitted to them.

Nor indeed could a remark like the following be left unanswered:

How Dr. Hutton, whose life, till he was mature, was spent in keeping a village school in Westmorland, could have sustained the office [of foreign secretary] without numberless offences against the habits of good society, it is difficult to conjecture; and his merits, as a mathematician, were commonplace.

Lands he could measure, terms and tides presage;
And even the saying ran, that he could gauge.

The inevitable reply appeared anonymously, savaging Banks's character and conduct as president in terms that would have been scarcely feasible when he was alive and were much less than decent now he was dead. No one seems to have doubted that it was the work of the mathematicians' usual spokesman Olinthus Gregory, although the article relied heavily on information that could only have come from Charles Hutton. It's possible to read it as Hutton's own last word in the affair, perhaps even with some use of text from the pamphlet he supposedly wrote and suppressed back in the 1780s. Banks was 'notoriously fond of farming, fond of grazing, fond of gardening'; he had written nothing but a 'little Essay on *blight*, and perhaps a diminutive disquisition or two on the manufactory of gooseberry-wine'; he 'evinced an absolute ignorance of several of the most interesting and useful sciences'. Malevolent, obstructive, bigoted, 'a good hater'. Bitter and ill advised all this was. But with Banks dead Hutton was indeed in a position to laugh last.

There was no further direct reply, and the attention of the Royal Society itself turned to the finding of a new president. Banks's own preference had been for William Wollaston, a shy man who had left medical practice for laboratory work in chemistry. Babbage, Herschel and George Peacock were keen on him, and they recruited various others to the cause, including Hutton. But Wollaston withdrew his candidacy, and attention turned to various others. Hutton received another canvassing visit, this time from one Humphry Davy.

A small man with a piercing glance and an engaging manner, the chemist and inventor was forty-two: barely half Hutton's age. His personal trajectory had some intriguing overlaps with the elderly mathematician's. The son of a Penzance woodcarver, he had worked for years on gases, following up the efforts of Priestley. He won the Copley Medal not for exalted pure research but for improvements in the smelly and nasty business of tanning. Most famously, he had been asked in 1815 to use his expertise in the chemistry of gases to do something about the problem of explosions in coal mines. He had travelled to Newcastle, met colliery viewers and visited collieries. There he had found that what the miners called 'firedamp' was methane; he invented a lamp that didn't ignite it, by enclosing the lamp in a mesh the flame was unable to cross. By 1816 Davy's safety lamp was in use in the North-East.

For all that, Banks had been in effect Davy's patron, and he was acceptable – and more than that – to at least some of the old guard at the Royal Society. The chance seemed good that he would secure the election. During his visit to Hutton he praised mathematics; said he was studying the subject himself with a view to under-standing the behaviour of elastic fluids. He was a good talker, a famed high-level populariser. Tolerant of specialisms and in favour, cautiously, of modernising the Society, his advent seemed to bode well for the mathematical party and indeed for the reintegration of British science more generally. Hutton was convinced, and agreed to support Davy for president.

Others did so too, in large enough numbers to make the contest an easy one. On 30 November 1820 Sir Humphry Davy, Bart., became the twenty-second president of the Royal Society.

∞

The next week Davy gave a discourse on the state of British science, paying particular attention to relations between the Royal Society and other scientific societies: especially the Astronomical Society. In the course of this he praised mathematics highly: 'the [highest] efforts of human intelligence'; 'abundant in the promise of new applications'; 'that sublime science which is as it were the animating principle of all the other sciences'.

He brought members of the Astronomical Society under his wing, supported reform of the *Nautical Almanac*, sought to extend scientific activity at the Board of Longitude. Babbage, Baily, Herschel, James Ivory and James South, astronomers and mathematicians all, were brought onto Council. Herschel was awarded the Copley Medal in 1821 for his astronomical work, and Davy took the opportunity to praise mathematics again in his speech on the occasion: 'There is certainly no branch of Science so calculated to awaken our admiration as the sublime or transcendental Geometry'; Herschel's work must be gratifying 'whether the importance of the subject be considered, or the glory that has been derived by the Society from the labours amongst those of its Members who have cultivated the higher branches of the Mathematics'. Hutton could almost have written it himself. Mathematicians who had hitherto resisted came into fellowship of the society: Francis Baily, Benjamin Gompertz, Thomas Leybourn,

Peter Barlow, Samuel Christie; Hutton himself signed the nominations for James Andrew, principal of the East India Company's college, and for George Rennie, engineer.

Newly confident, Hutton reopened the matter of the density of the earth one last time. Davy in his inaugural speech as president had specifically mentioned 'the grand question of universal gravitation', alluding to the work of Cavendish but unfortunately not to that of Hutton. Hutton was still convinced Cavendish's torsion balance result was wrong, and persuaded himself that the calculations involved were incorrect, suggesting that an assistant entrusted with them must have made mistakes. He reportedly asked a number of his friends to repeat the calculations, but on their refusal he determined to do it himself. Another prompt to action was an article in the French journal the *Connaissance des tems* that once again omitted mention of Hutton in connection with the subject. Hutton corresponded with the editor, the great mathematician Pierre-Simon de Laplace, in a pair of open letters, and was eventually rewarded with a full acknowledgement in the *Connaissance* in 1820. Meanwhile, he recalculated Cavendish's result and submitted it to the Royal Society: his first paper sent there in thirty-seven years. It was read in April 1821 and published in the *Philosophical Transactions* later that year.

Hutton was eighty-three, and the paper was an error of judgement. Its fairly obvious purpose was to draw attention, again, to Hutton's own role in the Schiehallion work and to bring renewed attention to the value for the density of the earth which resulted from it: now fudged up to 'very near five' times that of water by the use of Playfair's measurements of the densities of different rock strata found on the mountain. Hutton identified a number of 'errors' in Cavendish's work, but several of them were simple misreadings of the details of his paper: he took the large balls Cavendish used to be ten inches wide rather than twelve, and failed to notice an obvious printing error. By these means he was able to persuade himself – if not perhaps anyone else – that Cavendish's experiments, properly interpreted, yielded a value for the density of the earth of 5.31 times that of water, rather closer to that from the Schiehallion work than the value Cavendish himself had calculated.

∞

If the paper added nothing to Hutton's reputation, that reputation still stood very high indeed. A journalist suggested that 'perhaps no name can be mentioned, either ancient or modern, that has so successfully promoted those branches of Mathematical Knowledge, most conducive to the practical purposes of Life, as Doctor Hutton'. And the changes at the Royal Society had made it easier for friends and colleagues to express their admiration without fear for the consequences (it was said that in Banks's day at least one man's patronage had been withdrawn when it emerged that he had dedicated a book to Hutton).

In autumn 1821 a group of Hutton's associates met to devise a plan to honour their old friend. The core group included Gregory (of course), Francis Baily of the Astronomical Society (and now the Royal Society too), 'cousin' Catherine Hutton and others. They agreed to commission a bust of Hutton, to be paid for by subscription. The subscribers would pay one pound each – the amount was deliberately set fairly low to enable more people to contribute – and the noted Irish sculptor Sebastian Gahagan would do the work. They issued a prospectus and placed announcements in the newspapers, and they found themselves somewhat overwhelmed by the response. Subscriptions poured in from over 120 individuals and institutions, and the list began to read like a litany of the notable mathematicians and natural philosophers of the country, together with a large number of friends and former students of Hutton, all eager to express their gratitude and admiration: James Watt, Charles Babbage, Charles Burney, Benjamin Hobhouse,

Thomas Leybourn, John Rennie. The president of the Newcastle Lit and Phil subscribed, as did the Lit and Phil itself, and the Mayor and Corporation of Newcastle. So did institutions ranging from the famous Spitalfields Mathematical Society to the East India Company. Individual subscribers came from the Royal Artillery and the Royal Engineers, the Bengal Artillery and the Bombay Observatory, from Edinburgh, Aberdeen, and all around England. So much money was left over after paying the sculptor that the committee had a commemorative medal struck as well and presented to each subscriber. It showed Hutton in profile, with emblems of his two great scientific achievements: a balance to weigh the world, and a cannon suspended from a pendulum.

There was no real secrecy about the scheme, not least because Hutton read the newspapers like anyone else. Some of the subscribers, in fact, wrote directly to him rather than to the committee. He also had to sit for the bust.

Gahagan did his work well. There was anxiety among the subscribers that Hutton's character should be accurately caught, and the sculptor obliged with a portrait of the great mathematician in somewhat idealised old age. Vast forehead; not a wrinkle in sight. A rather determined mouth and the hint of a classical robe rather than contemporary dress (imagination could supply a laurel wreath if one was needed). Most thought the likeness very faithful, although there was some comment that it looked a bit gloomy: 'grave' was the word used. Hutton himself thought gravity a part of his character, and liked it.

On 21 September 1822 the committee waited on Hutton at Bedford Row to present him with his bust. It was an emotional occasion, the committee repeating phrases about their admiration and devotion to Hutton, their gratitude for his virtues, his talents and his long labours. Hutton himself, in a prepared response, spoke of 'an honour far beyond what I could have aspired to'. The Great Northern coalfield was likely in his mind, as well as the long years enduring Banks's disgust and disapprobation at the Royal Society.

The bust took up its station on a corner table in Bedford Row. Subscribers had been invited, at an extra cost of two guineas, to purchase casts of it for what the prospectus described – a little alarmingly – as 'veneration' at home; many did so.

One name on the subscribers' list caught Hutton's particular attention. John Scott, the Newcastle grammar school boy who had

The marble bust of Charles Hutton.

eloped with his sweetheart in 1772, had enjoyed a long and distinguished career in the law and was now, improbably enough, the Lord Chancellor of England. As Lord Eldon, his name headed the list, as befitted his rank, and Hutton wrote to thank him for subscribing. He recalled Scott and his wife, who had both been his pupils long ago, and said he remembered them 'with a sort of parental affection'.

There was one last, most gratifying mark of respect to come. In November of the same year – 1822 – the Council slate at the Royal Society included a name last seen there more than four decades before. Charles Hutton was once more elected an ordinary member of Council. No one could have missed the point of Hutton's rehabilitation; like the presentation of the bust, it was widely reported in the newspapers. It was a generous compliment to the grand old man of British mathematics, and a clear sign of Davy's intention to reconcile British mathematics and British science.

∞

Hutton never attended a meeting of Council in the 1822–3 season. It's uncertain, indeed, whether he attended meetings of the Royal Society at all during this period, though it's appealing to think he went along for the reading of his paper on the density of the earth in 1821. His health was poor. Respiratory disease, whatever its ultimate source, had dogged him since the 1780s. The winters of 1815 and 1816 were particularly hard on him; he spoke of feeling 'quite the old man' in the winter of 1815. From 1817 he had difficulty writing, and one day in January 1819 he was embarrassed to find he had confused the order of pages in writing a letter to Catherine. He suffered cold after cold; he was housebound for long periods.

Yet there were times when he seemed much better. Catherine remarked on the continuing vivacity of his letters, their freedom from any taint of old age. He remained mentally sharp, answering

Charles Hutton in old age.

correspondents punctiliously. From later in 1819, indeed, his health improved and he felt better than he had for years. In 1822 alone he saw through the press new editions of the *Guide*, the *Measurer* and the *Tables*: twelve hundred pages in total, dense with mathematics and numbers; a volume of work that would have exhausted men a third his age.

If social life was becoming less frenetic, he was no recluse, and the flow of visitors to Bedford Row showed no sign of drying up. One visitor in the spring of 1822 reported in some detail on Hutton's preoccupations and his manner. He seemed hale and in remarkably good spirits, telling the old tale of his relationship to Isaac Newton with some relish. He could read without spectacles (he claimed), though he was a little deaf. He called 1822 the happiest year of his life.

But Hutton was frail. In the winter of 1822 he caught a cold that brought back his lung complaint. Conversations with Gregory turned on his approaching death, and he took steps to put his affairs in order. Letters and manuscripts were labelled; some items were given to Gregory or set aside to be given to him. In January he made his will.

On Friday 24 January he received a letter from the Corporation of London, asking once again for an opinion about London Bridge, the curves that should be adopted for the arches in the new structure. Would it work, would it hold? Gregory visited the same day for a detailed conversation about the subject. The eighty-five-year-old 'expatiated with his usual perspicuity and accuracy upon the theory of arcuation, the relative advantages and disadvantages of different curves selected for the intrados, the most judicious construction of contering, &c . . .' He dictated a letter, and when he was tired Gregory agreed to visit again in a week.

Over Saturday and Sunday Hutton weakened, though he retained his mental faculties almost to the very end. By Sunday evening he was unconscious, and around four on Monday morning, 27 January 1823, he died.

He was buried the next day, in the churchyard at Charlton near Woolwich, where lay his daughter, his grandson and his wife. The churchyard faced open country; from it you could see down the long slope to the river, and beyond it towards the north.

Epilogue

'Dr. Hutton is gone where, we trust, all the labyrinths of the universe
will be revealed to him; leaving, to mathematicians, a name seldom
equalled for science, for utility, never; and, to his friends, the memory
of a character adding to that science an unwearied fund of knowledge and conversation, a cheerful and kind disposition, and the
simplicity of a child.'

– John MacCulloch, 1824

∞

Charles Hutton's was a long journey, in more ways than one, and
he had known triumph and despair in more than common measure.
The eighteenth century was an era of significant social mobility;
it was also an era of significant change, social, intellectual and
cultural. Hutton lived from the age of Alexander Pope to that of
Walter Scott and Ann Radcliffe; through the entire lifetimes of Jane
Austen and Percy Shelley.

He saw American independence, the French Revolution, and
much of what has come to be called the industrial revolution. Yet
even when all that is taken into account, he himself had been on
a long and a spectacular journey. By the time of his retirement and

even more so by the time of his death he was a national celebrity. His death was reported in almost every London newspaper, and his son received condolences from both the Duke of Wellington and the Lord Chancellor. People named their sons after him: Charles Hutton Potts (soldier), Charles Hutton Lear (artist), Charles Hutton Dowling (mathematician). The pages of *The Ladies' Diary* flowed with poetic tributes:

> This country shall entwine
> Th'immortal wreath his matchless toils have won;
> And Science breathe in other lays than mine.
> A requiem o'er her lost – her darling son!

The *Morning Post* called him 'one of the most successful promoters of useful science perhaps in any age or country', the *Literary Chronicle* a 'great man . . . dear to science, whose memory will long be revered'. Humphrey Davy reckoned him 'one of the most able mathematicians of his country and his age'. The *Quarterly Review* judged that his popularity 'promises to be as permanent as it is extensive'. Such has not, in fact, proved to be the case.

In the first instance Hutton's memory and his reputation were guarded by his family. Henry Hutton retired to Ireland where he died just four years after his father, in June 1827. Since the 1780s, his interest in antiquities had grown into an obsession, and his handsome pension from the Royal Artillery gave him leisure to indulge in the collection of material – notably on the architecture of Scottish church buildings – that he showed no real sign of arranging for publication. He left a mass of papers, and despite the efforts of Catherine Hutton and, at her instigation, Walter Scott, publication was not attempted; they languish today in – mostly – the Advocates' Library in Edinburgh: twelve unpublished and unpublishable volumes of notes, drawings and lists.

Isabella lived quietly on in London until her death in 1839. She eventually moved out of the house in Bedford Row to a smaller establishment, and kept in touch with the circle of her relations and friends and to some degree the friends of her father. Her correspondence with Catherine Hutton was a lively one, and there was even a small item in the newspapers about the enduring friendship of these ersatz cousins – who knew perfectly well that they were not related. Eleanor was widowed again and married for a third time, to a George Children; her later life is obscure, and she died in March 1850.

Margaret had once hoped that no grandchild of Charles Hutton would live out his life 'a meer ordinary being', and it is one of the great surprises of Hutton's story that of all the members of the next generation – that of his grandchildren – it was Charles Blacker Vignoles who most lived up to that challenge.

Vignoles, still in America, heard of his grandfather's death through the newspapers at the end of March. He felt a moment of indecision, or rather of determination to remain where he was. 'I do not expect this will cause any material change in my destiny, and not having at any time calculated upon any advantage, the circumstance is a matter of much regret, as I have wished to have met him in an independent state before he died.' But he quickly changed his mind and sailed home, leaving both his debtors and his creditors behind him.

Henry Hutton would never entirely forgive Vignoles for the way he had behaved, and his descendants preserved a memory of Vignoles's 'ingratitude' towards his uncle for generations (there was a letter to the press on the subject as late as 1889). But Isabella had always had a soft spot for the young man, and it took her a matter of hours to admit him back into his old room in Bedford Row. Both were quite eager to persuade themselves that the long rift had been merely a terrible misunderstanding.

Things could still be difficult; at first Isabella did not want to see Vignoles's wife and children, and there was something of a scene when they moved back to London and she received the impression they were going to come and install themselves in Bedford Row at once. In fact Isabella relented on that matter too, and in just a few weeks she allowed Mary and the children to move in with her. She was sixty, and abhorred both noise and unnecessary expense, but she seems to have reconciled herself to the presence of the young family and its demands (five-year-old Camilla was wont to practise the piano) for a time, before they moved on to an establishment of their own. Vignoles would name his fourth child Hutton 'in memory of my Grandfather & in respect to my Aunt Hutton'; another was called Isabella. This was all the reconciliation there would ever be with Charles Hutton, but there is perhaps at least a hint here that Vignoles had come to feel he, too, had behaved badly.

Vignoles's career took flight now he was out of the shadow of his famous grandfather. On the strength of his surveying work in America and his publications there, he established himself as a civil engineer. Olinthus Gregory helped, as did Thomas Leybourn, and the link with Charles Hutton did no harm; more than once Vignoles traded on the fact that he had learned mathematics with the great man. He worked on canals, bridges and – during the mania of the 1840s – railways; he built bridges and railways in Russia and Switzerland, Spain and Brazil, and more British lines than Brunel. The flat-bottomed shape of most railway rails – the 'Vignoles rail' – is his most enduring legacy. In 1841 he was the first Professor in Civil Engineering at University College London; in 1855 he was elected a Fellow of the Royal Society.

His domestic life was unhappy. The necessarily mobile life of a civil engineer was hard on his family, and the years of separation had already taken their toll on him and Mary. For many years Vignoles acted for the best according to his notions, in circumstances often very difficult and seldom entirely of his own making. But accusations of neglect and cruelty grew steadily wilder on both sides, and eventually Mary's language and behaviour became so erratic that she was placed in the care of a nurse. Her letters from this final period make pitiful reading; she died in 1834 of 'dropsy in the chest'.

Vignoles's own daughter Camilla seemed to threaten a repetition of a sorry pattern when she contracted a clandestine marriage to a Mr Croudace, for whom her father felt no enthusiasm ('amiable blockhead' was the phrase he used in a letter to Isabella).

A generation later Charles Hutton passed out of living memory. Eleanor's children have so far proved untraceable; Vignoles died in 1875 and his cousin, Henry's second son, in 1863, after a life in the church. The line of Charles Hutton's descendants continued on the one hand with Vignoles's children (five reached adulthood) and on the other with another Henry: Colonel Henry Francis Hutton. On the Vignoles side the line continues to this day.

Thus Hutton's family. What of his rumoured wealth? Hutton owned about forty-five thousand pounds at his death. Not an immense fortune, but close to a hundred times his annual pension, and undoubtedly a good deal for a pit lad. His will, written on 7 January 1823 when he had just three weeks to live, left the marble bust of himself to the Literary and Philosophical Society in Newcastle, where it still stands, looking benevolently across a pleasantly bustling reading room. It left everything else to his unmarried daughter Isabella. She had been a constant and faithful companion to her father for something like fifty years, and she had no means of support beside the income from her inheritance.

Nevertheless, some in the family resented the one-sided arrangement. Vignoles wrote bitterly about the matter to his wife, and it was rumoured as far away as Newcastle that Henry Hutton planned to dispute the will in Chancery, even that Isabella had offered to give up the principal in exchange for a modest annuity. It never came to that, but a trust was set up and in the first instance Isabella agreed that it would pay annuities to her brother and sister. She reckoned the remaining income was still twice what she needed, and it was understood that the surplus would be allowed to accumulate. In later years payments were also made to Henry's son. The trust's finances became rather complex; it invested in mortgages, and one large holder defaulted and was declared bankrupt, but the details are desperately obscure. Payments continued to be made, but Vignoles would state in 1839 that the trust now had barely enough value to be able to continue.

Vignoles was, it appears, the last surviving trustee, and my understanding is that he ultimately became the sole owner of the remaining capital and property, whatever they were worth. At this stage he passed on the Vanderbank portrait of Newton to the Royal Society; some other portraits, including one of Hutton and another of Isabella and Camilla, stayed in the Vignoles family. So did Vignoles's own voluminous diaries and correspondence, which eventually went to public libraries in Portsmouth and London.

One of the enduring mysteries about Charles Hutton – and it bears on the survival or otherwise of his reputation – is the destiny of his own papers. At a very conservative estimate Hutton must have written or received a letter a day from the mid-1760s to the mid-1810s. Even if he kept only the most interesting 10 per cent, that would still have made a collection of well over a thousand letters, including material from many of the biggest scientific names in Britain and Europe. And at one time, presumably, the letters were kept together in some sort of sequence: Hutton was a notoriously methodical man, and for some the secret of his success lay partly in the order with which he kept his private notes. As well as letters there were certainly some scientific manuscripts. There was even a diary, which Gregory had in his hands for part of 1823, 'and memoirs of his life and writings'.

The diary and memoirs are gone; of Hutton's letters only about 130 have turned up, and they are scattered, today, across more than thirty locations. His scientific papers are similarly scattered, and similarly few in number. What happened?

To begin with, neither the military installation at Woolwich nor its parent the Board of Ordnance, based at the Tower of London, was a conducive environment to the safekeeping of manuscripts and artefacts. Hutton's novel eprouvette was certainly destroyed in a fire at the Woolwich Repository in 1802, and there were further fires at Woolwich in 1873 and at the Tower in 1841, which would have affected any of his papers that remained there. His manuscript lectures on natural philosophy were said by Gregory to have been lost 'in a very extraordinary manner' around 1813, which sounds like a spectacular accident at Woolwich. His contract with the Stationers' Company obliged him to return to the company his almanac-related manuscripts when his work for them ended, and there is no sign that the company preserved them. Hutton himself is unlikely, I think, to have retained personal papers and letters that bore even obliquely on the failure of his first marriage or the illegitimacy of his youngest daughter.

On the other hand, Hutton certainly sorted and labelled at least some of his scientific manuscripts and some of his letters, and there is no doubt that he passed certain of them to Gregory during the 1810s, including writings on ballistics that were useful to Gregory in his own work on the subject. By the time Gregory was writing a long obituary of Hutton in early 1823 he had possession of several more scientific items, including the translation from Tartaglia and the notes on the history of algebra and on the ancient geometers Archimedes and Pappus, along with Hutton's collection of memorabilia from the 1784 affair, as well as at least a few letters, printed books, and even a pendulum. Some of this had been presented to him after Hutton's death, but our evidence fails to make it clear how much.

Gregory had the notion of working on some of the topics represented in the papers he received, particularly including ancient geometry, but in the event he did not, nor did he edit any of Hutton's work for posthumous publication. He also expected that Hutton's diary would appear in print, but it never did, despite some public comment on the matter. Gregory died in February 1841, and his books and papers were sold at auction. A few mathematicians, including the great Victorian mathematician Augustus De Morgan, made a point of attending and acquiring Hutton-related material, with the result that there are Hutton manuscripts in University College London and Trinity College Cambridge today. But the diary vanished without trace.

As for anything that didn't go to Gregory – including, I imagine, the bulk of Hutton's personal correspondence – the trail is still colder. A few items turned up in the hands of private collectors during the 1830s, but it's hard to say how they got there and harder to guess where they may have ended up in later decades. During his time in Bedford Row, Vignoles helped his aunt to dispose of some of the remaining books, and there are references in his journal to 'arranging papers', which could be those of Hutton; he would later state that he himself received only a 'few scraps' from Hutton's manuscripts. Elsewhere he elaborated that 'accident' had 'destroyed many' (he corrected himself to 'some') of his family's papers. I can believe in the few scraps, but my belief wavers as to the accident, which left Vignoles's as the only voice to be heard in the row between him and his grandfather. Whether by accident or design, then, it seems that the majority of Hutton's correspondence as well as his autobiographical writings – possibly together with a residue of scientific material – were probably destroyed in the decade or two after 1823. The material that passed through Gregory's sale is virtually all that is known today, apart from a hundred-odd letters to or from Hutton, very scattered.

Of physical reminders of Hutton's life we have, too, surprisingly little. The house in Newcastle where he was born succumbed to rebuilding in the nineteenth century; aptly enough, part of Newcastle University stands on the site today. The pit he worked in is covered by new housing, the location barely discernible. Woolwich Arsenal has been redeveloped again and again, although you can still see the Academy building where Hutton taught for so long. His own houses on Woolwich Common were pulled down during the nineteenth and twentieth centuries; his grave at Charlton was obliterated by new building. But the street, and the house, where he spent his last years and died are still standing, externally much as he knew them: Bedford Row, north of Holborn. Rows of smart Georgian town houses. A few trees. It was a fine winter day when I visited: Charles Hutton's front door; Charles Hutton's door scraper.

And Hutton's reputation, his popularity that 'promises to be as permanent as it is extensive' and his achievements that would always be remembered? What, indeed, had he achieved? On the military side, thirty-odd years of pressing the importance of mathematics in officer training had left a permanent mark on the Royal Artillery and the Royal Engineers, which were now, by several contemporary accounts, a versatile and supremely valuable body of men: 'Military tactics have been much benefited by his important labours, for it is by him that our artillery, and system of engineering have been brought to that perfection which they are universally admitted to possess.'

Through the example of what he did at Woolwich and its export to other institutions including the Royal Military College, through the appointment of personnel at Hutton's nod or at least sympathetic to his aims, and through the use of his textbooks, Hutton imparted at least a mathematical tone to the whole of British Army officer training, a tone that would not be wholly silenced during the Victorian period.

Davy reckoned Hutton had 'eminently contributed to awaken and keep alive that spirit of improvement among the military students which has so much exalted the character of the British officer, and which has been attended with such beneficial results to the country'. Hutton had indeed, with his colleagues at Woolwich, established a tradition of military improvement through scientific experiment that would cumulatively contribute much to British military success: indeed, had already done so. Woolwich, by the time of his retirement, had a well-established reputation as an experimental site of the 'utmost importance to the British nation', semi-public and at the public service, producing such innovations as shrapnel shells and artillery rockets.

On the scientific side, too, Hutton had pressed for the importance of mathematics and mathematical practice through forty years of exclusion from the Royal Society, and his voice had not in the end gone unheard. Admittedly Davy's reforms at the Society stalled, the active members feeling he was doing too little and others that he was doing too much, and history has judged him not to have made a success of the role of president. And by the time of William Whewell and Mary Somerville the British scientific community was (still) experiencing real anxiety about its lack of cohesion and – possibly consequent – lack of status. The term 'scientist', indeed, was coined in that context as part of a rallying cry for unity and common identity. As late as 1851 Charles Babbage could write that 'science in England is not a profession; its cultivators are scarcely recognised even as a class'.

But a process had been set in motion that could not now be stopped, and it would never again be possible for the Royal Society to sideline mathematics to quite the degree that had been countenanced under Joseph Banks. British culture in general was taking mathematics more and more seriously, both as a tool of industry and engineering and as part of the education of a gentleman; in

1799 the Cambridge Senate House announced that in order to obtain a degree you would henceforth need a 'competent knowledge of the first book of Euclid, Arithmetic, Vulgar and Decimal Fractions, Simple and Quadratic Equations', as well as other material.

Furthermore, around Hutton and under his patronage the mathematicians, mathematics teachers and mathematical practitioners of London and its surroundings had acquired a self-conscious identity that issued forth in such tangible projects as the founding of the Institution of Civil Engineers in 1818 and of the Astronomical Society of London – later the Royal Astronomical Society – in 1820, and that beyond a doubt contributed much to British science and engineering during the crucial years of the industrial revolution. (Hutton became a Fellow of the Royal Astronomical Society less than a year before his death; he donated a few books and a copy of his portrait of Newton to the Society, but played no active part in its affairs.) Hutton's colleague Thomas Leybourn produced a periodical called *The Mathematical Repository* from 1799 to 1833, partly fulfilling the old promise of a rival to the *Philosophical Transactions* on behalf of the mathematical party. Meanwhile, mathematical papers appeared in the *Transactions* of the Cambridge Philosophical Society from 1821 and in the *Cambridge Mathematical Journal* from 1837. Hutton's position as a sort of father figure in this group was emphasised from time to time by that most public declaration of intellectual allegiance, the dedication of a book to him: textbooks, translations and even poems were over the years placed under his notional 'protection'. The mathematical community would remain self-conscious, poorly organised, and sometimes rather at odds with the different values of the Royal Society right up to the 1860s, when at last the London Mathematical Society was formed.

At the same time, the mathematical culture to which Hutton had devoted his life had begun in his lifetime a process of change, moving it towards professionalisation and towards the adoption of French notation and conceptual vocabulary. The genial amateurism of *The Ladies' Diary* and its sisters, for so long a prominent feature of British mathematical culture, did not last. The *Diary* merged with *The Gentleman's Diary* in 1841 and closed in 1871, not before adopting French notation in place of British. Hutton himself was the author of a number of small mathematical novelties – tiny geometrical theorems and algebraic proofs, computational tricks and ways of calculating trigonometrical ratios – that were typical of that world, but did not entitle him to fame or commemoration in the new world of professional mathematics where conceptual innovation was prized above all. Equally, as editor of the *Diary* he had overseen the publication of a mass of tiny mathematical discoveries that in the long run were scarcely remembered. His perhaps more significant innovation in the theory of bridges, what he called the 'arch of equilibration', passed into the common stock of knowledge without his name remaining attached to it.

Meanwhile Hutton's two key scientific pieces of work – those commemorated on the medal: weighing the world and the force of fired gunpowder – went the way of all scientific ideas: eventually they were ousted by better ones. In the long run the torsion balance experiments of Cavendish were repeated with increasing success, and it came to be accepted that mountain-based measurements of the earth's density such as those Hutton had worked on could not be made similarly satisfactory, due to the difficulty of surveying enough territory and of guessing at the rock strata that it was impossible to see. It also came to be accepted that Hutton's own laborious calculations were marred by the coarseness of the grid of points he had used close to the observation stations, leading to a substantial error in the final result. Fine science in its day, in other words, but like all fine science destined to be superseded by finer.

As to Hutton's work on ballistics – on the force of gunpowder and the nature of air resistance. It was persistently asserted, including by Hutton himself, that the work had led to tangible changes in artillery design and practice: reduced windage, smaller charges, and the introduction of shorter cannon under the name of carronades. But such claims proved hard to substantiate in detail, and Hutton's name did not come to be decisively associated with any particular change in artillery. Perhaps rightly.

On the other hand, Hutton's experimentally derived formula for air resistance did come into wide use, and in the early twentieth century it was still possible for an informed observer to state that it was used 'perhaps more than all others combined'; it was still found in many up-to-date textbooks. But it had come to be called 'Unwin's formula', for reasons already obscure by that date: Professor Unwin it seems had done no more than quote Hutton, adding his approval.

Thus, even by the time of Hutton's death well-meaning admirers could write that his work was mainly in 'improving and simplifying', that he lacked originality, that he was as remarkable for his industry as for the intellectual content of his writings. It wasn't true, but it was how he would come to be remembered. Friends, colleagues and disciples would try in vain to correct the picture. Gregory and others insisted that two at least of his scientific papers – on gunpowder and the density of the earth – were 'the most useful and important that, perhaps, had been communicated since the chair of that learned Institution was filled by Sir Isaac Newton'. His was the minority view.

More direct attempts to tell Hutton's story also came to nothing. There were obituaries in national and provincial newspapers, many copied from one another; perhaps amounting to ten substantially distinct accounts of Hutton's life. Brief biographies appeared in at least four volumes of local Newcastle history later in the nineteenth century, alongside two short 'lives' that had appeared during his lifetime. Both Knight's *Cyclopedia* in 1867 and in 1891 the *Dictionary of National Biography* included entries about Hutton.

Together with the 'tribute of respect' associated with the bust presentation this gives us quite a lot of Hutton biographising from the period when he was within – or close to – living memory; there are also a few reminiscences in manuscript, collected by Newcastle historians probably at the behest of his correspondent John Bruce. But all this added up to a chaotic, not a coherent account of his life. London-based writers doubted Hutton had ever been in the coal pits; Newcastle-based ones garbled the events of 1784. No one said what had happened to the first Mrs Hutton, nor was any very clear account printed of how Charles Blacker Vignoles fitted in to Hutton's life.

At the time of writing his obituary, Gregory intended to compose a longer account of Hutton's life, perhaps even a book-length biography, based on his own diary. That's the last we hear of either Hutton's diary or the life by Gregory. Gregory's illness and death may be to blame: maybe material was suppressed by Vignoles; a later story said Henry quashed the biography because he didn't want it known his father had worked in the coal pits. No version is certain. In Gateshead, the antiquarian John Bell got as far as printing a title page for 'Collections relative to Charles Hutton'; but he printed nothing more.

It wasn't all failure, though. Hutton's colleagues did most effectively succeed in keeping his name before the public through republication of his books. The *Guide* received American and Scottish editions; for a period there were not two but three rival versions of the book on the market in Britain, and the very last went on sale in 1867. The *Measurer* and the *Recreations* were reissued, the latter by Edward Riddle, a protégé and disciple who also wrote a short life of Hutton as a preface. Editors of these posthumous versions differed in their attitudes, some prizing fidelity to Hutton's thought more than others. The *Course* in particular came to reflect the transformation in British mathematics, with French-style notation adopted in its later editions.

The *Course* also went to America, and Robert Adrain both taught from it and edited an American edition in New York, printed five times over the next decade and a half. There were British editions up to 1860; it was used at West Point until at least 1825 and at the Royal Military College into the 1830s. The *Course* was a particular hit, it seems, in British India, where the Bengal Artillery and the Bombay Engineers included men who had personal connections with Hutton. Translations were arranged into Gujarati, Arabic, Sanskrit and Marat'ha at Bombay and Calcutta during the 1830s. Together with an Urdu version of the *Conics* printed in Delhi in 1848 and a Japanese version of his work on bookkeeping, this gave Hutton an international reach probably exceeded only by Euclid among mathematical writers at the time. There was a period in the 1830s when the sun scarcely set on his textbooks. A list of authors owing debts to them would be a conspectus of virtually everyone who published on mathematics in the Victorian world, and many more besides. John Henry Newman owned a copy of the *Compendious Measurer*; Karl Marx cited Hutton's *Course*.

Finally the *Tables*, a work that went out of date less easily and was less affected by changes in mathematical style, went through half a dozen editions during the nineteenth century after Hutton's death; a writer in the 1891 *Dictionary of National Biography* reckoned the introduction still valuable as 'an interesting and learned history of logarithmic work'. The last edition of the tables was issued in 1894. When after a few years it fell out of print, the voice of Charles Hutton – pit boy, professor and unequalled spokesman for mathematics – at last fell silent.

Acknowledgements

My thanks, as always, to Jessica and to my parents for supporting this project in many and various ways; also to Aileen Mooney for invaluable assistance in securing the funding which made the research possible. Among my colleagues, Jacqueline Reiter and Anna Marie Roos were so kind as to read the book in draft and suggest valuable improvements: they bear no responsibility for its remaining defects. I am grateful to the participants at the conference on Charles Hutton in Oxford in December 2015, and especially to Emily Winterburn and Shelley Costa who helped to refine my thinking about Charles Hutton in the context of editing his collected correspondence. My special thanks to Camilla Barnes, Jim Campbell and John House for their help with Figure 13, and to John Vignoles for detailed information about the Vignoles family. Felicity Bryan and Arabella Pike believed in the book at a crucial stage and shepherded it from draft to publication; my thanks to them, and to the staff at Felicity Bryan Associates and William Collins who worked on the book.

My thanks also to the Arts and Humanities Research Council, which funded the research for this book; to All Souls College and the History Faculty, University of Oxford, which provided support during research and writing; and to the following libraries and archives and their staff: Birmingham Archives, Heritage and Photography Service; the Bodleian Library, University of Oxford; the British Library; Cambridge University Library; Durham Record Office; Firepower: the Royal Artillery Museum; the Harry Ransom Center, University of Texas; the Lewis Walpole Library, Yale University; the London Metropolitan Archive; the National Archives, Kew; Newcastle Literary and Philosophical Society; the Northumberland Collections Service, Woodhorn; Plymouth and West Devon Record Office; Portsmouth History Centre; the Royal Society Archive; the Sandhurst Collection; Senate House Library, London University; the Smithsonian Institution; the Library of Trinity College, Cambridge; the Tyne and Wear Archive; the University of Cincinnati Library; and the Wellcome Library.

Select Bibliography

The reader wishing to delve more deeply into Hutton and his world will find in the following books some of the more readable accounts of the topics and people concerned. A complete bibliography of the works cited in the notes can be found online at www.benjaminwardhaugh.co.uk, including a catalogue of Charles Hutton's surviving manuscripts and a list of his published works. His surviving letters are edited in Benjamin Wardhaugh, *The Correspondence of Charles Hutton (1737–1823): mathematical networks in Georgian Britain* (Oxford, 2017).

Carman, W.Y. and Michael Roffe, *The Royal Artillery* (Reading, 1973).

Carter, Harold B., *Sir Joseph Banks, 1743–1820* (London, 1988).

Colley, Linda, *Britons: forging the nation* (London, 2003).

Crookston, Peter, *The Pitmen's Requiem* (Newcastle, 2010).

Freese, Barbara, *Coal: a human history* (Cambridge, MA, 2003).

Graham, C.A.L., *The Story of the Royal Regiment of Artillery* (Woolwich, 1928, 4th edn 1983).

Higgitt, Rebekah, ed., *Maskelyne: Astronomer Royal* (London, 2014).

Hogg, O.F.G., *Artillery: its origin, heyday and decline* (London, 1970).

Howse, Derek, *Nevil Maskelyne: the seaman's astronomer* (Cambridge, 1989).

Hughes, B.P., *British Smooth-Bore Artillery: the muzzle loading artillery of the 18th and 19th centuries* (London, 1969).

Jungnickel, Christa and Russell McCormmach, *Cavendish: the experimental life* (Lewisburg, PA, 1999).

Lewis, C.L.E. and S.J. Knell, *The Making of the Geological Society of London* (London, 2009).

Moffat, Alistair and George Rosie, *Tyneside: a history of Newcastle and Gateshead from earliest times* (Edinburgh, 2005).

O'Brian, Patrick, *Joseph Banks: a Life* (London, 1987).

Seymour, W.A. and J.H. Andrews, *A History of the Ordnance Survey* (Folkestone, 1980).

Shepperd, G.A., *Sandhurst: the Royal Military Academy Sandhurst and its predecessors* (London, 1980).

Smith, Ken and Jean Smith, *The Great Northern Miners* (Newcastle, 2008).

Southey, Roz, Margaret Maddison and David Hughes, *The Ingenious Mr Avison: making music and money in eighteenth-century Newcastle* (Newcastle, 2009).

Uglow, Jenny, *Nature's Engraver: a life of Thomas Bewick* (London, 2006).

Vignoles, Keith Hutton, *Charles Blacker Vignoles, romantic engineer* (Cambridge, 1982).

Wardhaugh, Benjamin, *Poor Robin's Prophecies: a curious almanac, and the everyday mathematics of Georgian Britain* (Oxford, 2012).

Warwick, Andrew, *Masters of Theory: Cambridge and the rise of mathematical physics* (Chicago, 2003).

Notes

1 Out of the Pit

2 *Tyne river, running rough or smooth*: Ken and Jean Smith, *The Great Northern Miners* (Newcastle, 2008), 3. See also Eneas Mackenzie, *A Descriptive and Historical Account of Newcastle-upon-Tyne* (Newcastle, 1827); P.M. Horsley, *Eighteenth-Century Newcastle* (Newcastle, 1971); C.M. Fraser and Kenneth Emsley, *Tyneside* (Newton Abbot, 1973); and Alistair Moffat and George Rosie, *Tyneside: a history of Newcastle and Gateshead from earliest times* (Edinburgh, 2005).

2 *Charles Hutton's parents*: Genealogical information is collated from family-search.org, freereg.org.uk, ancestry.com, and findmypast.com, as well as from the inscription from Hutton's gravestone (now destroyed) transcribed in Leonard Morgan May, *Charlton: Near Woolwich, Kent. Full and complete copies of all the inscriptions in the old parish church and churchyard* (London, 1908), 69.

2 *a 'viewer' in the collieries*: On the work of coal viewers see *The Compleat Collier* (Newcastle, 1730), 18–19; also *A Pitman's Notebook . . . the diary of Edward Smith, transcribed, edited and annotated by T. Robertson* (Newcastle, 1970); and *The Hatchett Diary*, ed. A. Raistrick (Truro, 1967).

3 *High-end viewers*: See John Hatcher et al., *The History of the British Coal Industry* (5 vols, Oxford, 1984–1993), vol. 2, 58–62; and J.R. Harris, 'Skills, Coal and British Industry in the Eighteenth Century', *The Journal of the Historical Association* 61 (1976), 170.

3 *a land steward for Lord Ravenscroft*: Olinthus Gregory, 'Brief Memoir of the Life and Writings of Charles Hutton', *Imperial Magazine* 5 (March 1823 [obituary dated 1 February]), 201–27 at 201. Gregory was probably relying on Hutton's own reminiscences, and although John Bruce ridiculed the claim

in *A Memoir of Charles Hutton* (Newcastle, 1823), 3, he provided no evidence that it was false.

3   *this 'low' dwelling*: Mackenzie, *Historical Account of Newcastle*, 557.

3   *compared with the cottages of the miners*: Edward Smith, *A Pitman's Notebook*, 6; Peter Crookston, *The Pitmen's Requiem* (Newcastle, 2010), 22.

4   *Ladies liked little Charles*: 'Charles Hutton' in *Public Characters* (10 vols, London, 1799–1809), vol. 2, 97–123, at 98.

4   *actual hangings were a rarity*: Moffat and Rosie, *Tyneside*, 207.

4   *an old Scots woman*: Bruce, *Memoir*, 4.

5   *an 'overman' in one of the collieries*: Bruce, *Memoir*, 3; also Woodhorn, SANT/BEQ/26/1/7/77 (notes of Thomas Wilson, 28 March 1822).

5   *Hundreds of thousands of tons*: The History of the British Coal Industry vol. 2, 26.

5   *The so-called Grand Alliance*: The History of the British Coal Industry vol. 2, 41, 61, 157.

6   *The members of the Grand Alliance were technophiles*: The History of the British Coal Industry vol. 2, plate 7 and pp. 101, 122; also William Fordyce, *A History of Coal . . .* (London, 1860), 82; Robert Edington, *A Treatise on the Coal Trade* (London, 1813), 115; and *A Pitman's Notebook*, 35.

6   *housing for the underground workers*: The History of the British Coal Industry vol. 2, 434–40.

7   *massive revelry at weddings*: See Edward Chicken, *The Collier's Wedding* (Newcastle, 1764); compare *The Colliers' Rant* (Newcastle, 1740) and Thomas Wilson, *The Pitman's Pay* (Gateshead, 1843).

7   *Middle-class visitors*: History of the British Coal Industry, 434–40; *The Journal of the Rev. John Wesley*, ed. Nehemiah Curnock and John Telford (8 vols, London, 1909–1916), vol. 3, 13 (arrival in Newcastle on Thursday 27 May 1742).

7   *operating a trapdoor in the pits*: Obituary of Hutton in the *Newcastle Chronicle*, 31 January 1823. On the conditions of work see *The Pitmen's Requiem*, 146–7; Smith, *Great Northern Miners*, 19; Barbara Freese, *Coal: a human history* (Cambridge, MA, 2003), 77; and Wilson, *Pitman's Pay*, x.

8   *Firedamp and chokedamp*: The Hatchett Diary, 81; *The Pitmen's Requiem*, 30; see also *The Compleat Collier*, 9 and History of the British Coal Industry, vol. 2, 420.

8   *asking the newspapers not to print reports of pit explosions*: Freese, *Coal*, 51–2, citing the *Newcastle Journal* for 1767.

8   *You just clung on*: History of the British Coal Industry, vol. 2, 103; James Everett, *The Wall's End Miner* (London, 1835), 48.

8   *Eventually that would happen to Francis Frame*: 'A Memoir of Charles Hutton', *Newcastle Magazine* (June 1823), 298–311 at p. 299.

8 *your chance of dying in an accident*: History of the British Coal Industry, vol. 2, 419.

8 *the exploitation of children in the pits*: Commission for Inquiring into the Employment and Condition of Children in Mines and Manufactories, *First Report of the Commissioners* (London, 1842); also Sara Horrell and Jane Humphries, 'Old Questions, New Data, and Alternative Perspectives: the standard of living of families in the industrial revolution', *Journal of Economic History* 52 (1992), 849–80 and Sara Horrell and Jane Humphries, '"The Exploitation of Little Children": children's work and the family economy in the British industrial revolution', *Explorations in Economic History* 32 (1995), 485–516.

9 *There were schools*: History of the British Coal Industry, vol. 2, 438.

9 *Hutton injured his arm*: Bruce, Memoir, 3 is the fullest of several accounts in the obituaries.

9 *A schoolteacher named Robson*: Bruce, Memoir, 4; also *Public Characters*, vol. 2, 98.

9 *a new, locally produced, grammar book*: Anne Fisher, *New Grammar* (3rd edition: Newcastle, 1753).

10 *He was fond of the so-called 'border ballads'*: Public Characters, vol. 2, 100–01; also Gregory, 'Memoir', 201.

10 *the southward march of the Bonnie Prince*: Moffat and Rosie, Tyneside, 203–5.

11 *He preached in the fields*: The Journal of the Rev. John Wesley, vol. 3, pp.14, 59, 67–8, 80, 110–11, 140, 165; also Moffat and Rosie, Tyneside, 202 and Rupert E. Davies, *Methodism* (London, 1976), 59–60.

11 *reinventing yourself*: See Misty G. Anderson, *Imagining Methodism in Eighteenth-century Britain: enthusiasm, belief, and the borders of the self* (Baltimore and London, 2012), *passim*.

12 *anecdotes . . . concerned with his piety*: these all appeared first in *Public Characters*.

12 *'Never be unemployed for a moment'*: Davies, Methodism, 70; see also John Wesley, *The Character of a Methodist* (Newcastle, 1743).

12 *His teacher, Jonathan Ivison*: Bruce, Memoir, 4; also information from theclergydatabase.org.uk.

13 *envied by other students*: Public Characters, vol. 2, 98.

13 *Hutton acted as Ivison's assistant*: 'A Memoir of Charles Hutton', *Newcastle Magazine*, 300.

14 *'putting'*: The History of the British Coal Industry, vol. 2, 347–8; Smith, *Great Northern Miners*, 18; Compleat Collier, 15–16; Fordyce, *A History of Coal*, 32.

14 *a pay bill from the Long Benton colliery*: Bruce, Memoir, 5. Bruce discusses the evidence, which he appears to have seen; the relevant records are catalogued

at the Tyne and Wear Archive, but can no longer be located (personal communication from Alyson Pigott at Tyne and Wear Archives and Museums, 28 Apr 2015). Compare the probably independent report in Woodhorn, SANT/BEQ/26/1/7/77 (notes of Thomas Wilson, 28 Mar 1822): a Mr Kirkley was said to be 'certain that Hutton wrought in old Benton pits'.

14   *Put one leg through a loop in the rope*: History of the British Coal Industry, vol. 2, 103. On the physical appearance inside the shafts see Thomas Hair, *A Series of Views of the Collieries in the Counties of Northumberland and Durham* (1844), 7.

14   *Strip to the 'buff'*: Moffat and Rosie, *Tyneside*, 189–90: quoting E.A. Rymer, *The Martyrdom of the Mine*.

15   *vertical cuts from the top of the seam to the bottom*: History of the British Coal Industry, vol. 2, 91; also Wilson, *The Pitman's Pay*, xi and Smith, *Great Northern Miners*, 17.

15   *the elite of the pit*: E. Neville Williams, *Life in Georgian England* (1962), 99.

16   *The stars are twinkling in the sky*: Joseph Skipsey, *Songs and Lyrics* (London, 1892), 9.

16   *you shouldn't ever wish for more*: see John Wesley, *Instructions for Children* (Newcastle, 1746), 14; cf. 35.

17   *'taken from school'*: 'A Memoir of Charles Hutton', *Newcastle Magazine*, 299.

17   *He was 'laid idle' one day*: Northumberland Collections Service, SANT/BEQ/26/1/7/77 and SANT/BEQ/26/1/8/584, reporting two separate reminiscences of the incident by other miners.

2 Teacher of Mathematics

21   *lots of one-on-one teaching*: See Nerida Ellerton and M.A. (Ken) Clements, *Rewriting the History of School Mathematics in North America 1607–1861: the central role of cyphering books* (Dordrecht, 2012), 19–33; information also from presentations by Nerida Ellerton and Ken Clements at All Souls College, Oxford, December 2014 and December 2015.

21   *The school at High Heaton ran*: Woodhorn, SANT/BEQ/26/1/8/584: Letter of S. Barrass to Thomas Wilson, 1825.

21   *the schoolroom proved too small*: Bruce, *Memoir*, 6; John Sykes, *Local Records or Historical Register of Remarkable Events* (Newcastle 1833), vol. 1, 49.

22   *'Like most other young men of a liberal education . . .'*: Memoirs of Dr. Joseph Priestley, 24.

22   *Elementary teaching in particular*: Victor E. Neuburg, *Popular Education in 18th Century England* (London, 1971), 18.

22 *pay only slightly above that of a labourer*: Neuburg, *Popular Education*, Appendix 1; F.J.G. Robinson, 'Trends in Education in Northern England in the Eighteenth Century: a biographical study' (unpublished Ph.D. dissertation, Newcastle, 1972), 46.

22 *There was no teacher training*: Neuburg, *Popular Education*, 37.

22 *the boast of Dr Parr of Harrow*: Williams, *Life in Georgian England*, 36.

22 *Poor Reuben Dixon has the noisiest School . . .*: George Crabbe, *The Borough* (London, 1810), quoted in Neuburg, *Popular Education*, 20.

22 *'kept up the most rigid order'. . . carried 'his severity too far'*: Woodhorn, SANT/BEQ/26/1/7/77 (notes of Thomas Wilson, 28 March 1822).

22 *'assumed a degree of importance'*: Woodhorn, SANT/BEQ/26/1/7/77 (notes of Thomas Wilson, 28 March 1822).

23 *'his friends, who would have supported him'*: Woodhorn, SANT/BEQ/26/1/8/584: Letter of S. Barrass to Thomas Wilson, 1825.

23 *He read all he could*: Gregory, 'Memoir', 202; *Public characters*, vol. 2, 99, 101.

23 *'Newton's works and the works of his contemporaries'*: all these are cited in Hutton's *Guide* or his *Mensuration*.

23 *attended classes given by a Mr Hugh James*: Mackenzie, *Historical Account of Newcastle*, 557; Bruce, *Memoir*, 6.

23 *his mother feared for his health*: Bruce, *Memoir*, 6.

23 *his mother that died*: Parish register of St Andrew, Newcastle, 17 March 1760 (burial of Eleanor Frame), accessed via www.findmypast.co.uk; Bruce, *Memoir*, 8.

24 *he returned to St Andrew's to be married*: Durham Record Office, DDR/EJ/MLA 1/1760/70/1 and 70A: marriage licence and bond for Charles Hutton and Isabella Hutton, 7 April 1760.

24 *Soon there was a son*: 'Lieut.-Gen. Hutton', *The Gentleman's Magazine* (December 1827), 561–2; Charles Rogers, 'Register of the Collegiate Church of Crail, Fifeshire. With Historical Remarks', *Transactions of the Royal Historical Society* 6 (1877), 324–94 at 329; letter of P.J. Anderson in *Notes and Queries*, series 11, vol. 2 (1910), 347 (stating that the name *George Henry* – frequently given in print – is incorrect); Niccolò Guicciardini, 'Hutton, Charles (1737–1823), mathematician' in *ODNB*, incorporating 'George Henry Hutton'.

24 *TO BE OPENED On Monday, April 14th, 1760*: Quoted in Bruce, *Memoir*, 50.

24 *his fees were twice those asked by his rivals*: *Public Characters*, vol. 2, 102.

26 *Friends advised him to promise less*: Bruce, *Memoir*, 7.

26 *northern parents had a reputation*: Robinson, 'Trends in Education', 33.

26 *James Cook*: Moffat and Rosie, *Tyneside*, 193.

26 *Newcastle became England's best-educated city*: Neuburg, *Popular Education*, 79.

26 *at the Newcastle ironworks*: Neuburg, *Popular Education*, 80.

26 *Teaching mathematics at this level*: John Denniss, *Figuring It Out: children's arithmetical manuscripts, 1680–1880* (Oxford, 2012); also Hutton, *Guide*, *passim*.

26 *Hutton had his own special method*: Hutton, *Guide*, v.

27 *how many Flemish guilders will I buy*: Hutton, *Guide*, 100 (the answer is 1839 guilders, 2 stuivers, 11 5/6 penning: there were 16 penning in a stuiver and 20 stuivers in a guilder).

27 *If eight yards of cloth cost twenty-four shillings*: Hutton, *Guide*, 35.

27 *In how many days will eight men finish*: Hutton, *Guide*, 36.

27 *'Back-Row'*: Announcement by N. Stewart, Dancing-Master in *Newcastle Chronicle*, 24 March 1764; Advertisement for the sale of a colliery in *Newcastle Chronicle*, 11 May 1765; Bruce, *Memoir*, 10.

27 *selling tickets for the man's public balls*: Advertisement for a public ball in *Newcastle Chronicle*, 23 February 1765.

27 *a triangular prism 'is something like a hat box'*: Charles Hutton, *A Treatise on Mensuration, both in theory and practice* (Newcastle and London, 1770), 139.

27 *'soon conscious of his great abilities'*: Woodhorn, SANT/BEQ/26/1/7/77 (notes of Thomas Wilson, 28 March 1822).

28 *'a very modest, shy man'*: Letter from Isaac Cookson Esq. to Mr Thomas Hodgson, printed in Bruce, *Memoir*, 15.

28 *knocking a boy down in the street*: Woodhorn, SANT/BEQ/26/1/7/77 (notes of Thomas Wilson, 28 March 1822) and Woodhorn, SANT/BEQ/26/1/8/584: Letter of S. Barrass to Thomas Wilson, 1825.

28 *three more children*: Register of Hanover Square Presbyterian Chapel, Newcastle, 23 May 1762 (Isabella) and 25 April 1769 (Eleanor), accessed via www.freereg.org.uk and ancestry.co.uk. For Camilla see 'Deaths' in *The Gentleman's Magazine* (October 1794), 957 and Guicciardini, 'Hutton, Charles' in *ODNB*.

28 *The chapel there was established in 1727*: Mackenzie, *Historical Account of Newcastle*, 374.

28 *one report says he wrote sermons*: *Public Characters*, 100; also William Wood Stamp, *The Orphan-House of Wesley; with notices of early Methodism in Newcastle-upon-Tyne, and its vicinity* (1863), 245.

28 *a reputation for Unitarianism*: A.-H. Maehle, 'Clark, John (bap. 1744, d. 1805), physician' in *ODNB*; http://www.ukunitarians.org.uk/newcastleupon-tyne/history.htm; Constance Mary Fraser and Kenneth Emsley, *Tyneside* (Newton Abbot, 1973), 75; Moffat and Rosie, *Tyneside*, 215.

28 *questioned even . . . traditionally core doctrines*: *Memoirs of Dr. Joseph Priestley*, 23; Isabel Rivers and David L. Wykes, *Joseph Priestley, scientist, philosopher, and theologian* (Oxford, 2008), 190.

29 *Probate courts, marriage, schools and universities*: Rivers and Wykes, *Joseph Priestley*, 5, 11.

29 *a final move to Westgate Street*: Woodhorn, SANT/BEQ/26/1/7/77 (notes of Thomas Wilson, 28 March 1822); J.R. Boyle, *Vestiges of Old Newcastle and Gateshead* (Newcastle, 1890), 152.

29 *quite the elegant Georgian pile*: The house is described in an advertisement in the *Newcastle Courant*, 5 June 1773.

29 *The old town walls*: Moffat and Rosie, *Tyneside*, 210; Charles Hutton, *A Plan of Newcastle Upon Tyne and Gateshead, taken from an accurate survey, finished in the year 1770* (Newcastle, 1770, 1772).

29 *tall elegant buildings and wide open spaces*: Tyne and Wear Museums, TWCMS: 2003.1007 (print of 'Newcastle-upon-Tyne from the South'); TWCMS: G5272 (John H. Wilson, 'St Nicholas Church, Newcastle upon Tyne').

29 *three hundred street lamps and a well-organised night watch*: Hutton, *Plan*.

30 *Subscription concerts*: Roz Southey, Margaret Maddison and David Hughes, *The Ingenious Mr Avison: making music and money in eighteenth-century Newcastle* (Newcastle, 2009), 34–7, 91 and *passim*.

30 *A literary club; theatres*: Southey et al., *The Ingenious Mr Avison*, 47; Moffat and Rosie, *Tyneside*, 199.

30 *Scientific lectures*: Peter and Ruth Wallis, *Mathematical Tradition in the North of England* (Durham, 1991), 22–3; N.A. Hans, *New Trends in Education in the Eighteenth Century* (London, 1951), 110; Robinson, 'Trends in Education', 287–8, 302–20.

30 *The sale of coal was so lucrative*: Hutton, *Plan*.

31 *William Emerson*: William Emerson, *Tracts* (new edition, London, 1793), i–xxii.

31 *John Fryer*: Bruce, *Memoir*, 13; Mackenzie, *Historical Account of Newcastle*, 558.

31 *The school had its own separate entrance*: Advertisement for the let of the school in *Newcastle Courant*, 5 June 1773.

31 *'Youth are qualified for the Army, Navy, Counting-house'*: Advertisement in the *Newcastle Courant*, 7 April 1770.

31 *Newcastle Grammar School took to sending its students to Hutton*: Bruce, *Memoir*, 10.

31 *private academies did mathematics and science better*: Wallis, *Mathematical Tradition*, 16.

32 *Hutton's 'manners, as well as his talents'*: Charles Hutton, ed. Edward Riddle, *Recreations in Mathematics and Natural Philosophy* (new edition, London, 1840), vii.

32 *Robert Shaftoe*: Woodhorn, SANT/BEQ/26/1/8/584: Letter of S. Barrass to Thomas Wilson, 1825.

32 *he took to attending the lessons himself*: Bruce, *Memoir*, 7.

32 *a reading knowledge of French, Latin, Italian and German*: A Catalogue of the entire, extensive and very rare mathematical library of Charles Hutton, *L.L.D.* [London, 1816], *passim*; letter of Charles Hutton to Robert Harrison, 13 Jan 1779 in Sidney Melmore, 'Some Letters from Charles Hutton to Robert Harrison', *The Mathematical Gazette* 30 (1946), 71–81.

32 *public lectures in the subject*: Advertisement in the *Newcastle Courant*, 25 April 1772.

32 *taught mathematics to other schoolteachers*: Advertisement in the *Newcastle Courant*, 27 December 1766.

32 *external lecturers began to use it*: Charles Hutton, *Tracts on Mathematical and Philosophical Subjects* (3 vols: London, 1812), vol. 3, p. 379.

32 *Caleb Rotherham*: Hans, *New Trends*, 110.

33 *John Scott*: 'Memoir of the Right Hon. John Scott, Earl of Eldon', *Imperial Magazine* (July 1827), 593–600 at 593.

33 *Both went to Hutton*: Lord Eldon to General Hutton, 3 February 1823: letter printed in Bruce, *Memoir*, 47. See also Moffat and Rosie, *Tyneside*, 207, and Wilson Hepple, *The Elopement of Bessie Surtees* (oil on canvas, 1890).

3 Author

35 *The majority of mathematics teachers*: Hans, *New Trends*, 68–9.

35 *a dozen or so monthly or annual magazines*: Raymond Clare Archibald, 'Notes on Some Minor English Mathematical Serials', *The Mathematical Gazette* 14 (1929), 379–400; also Olaf Pedersen, 'The "Philomath" of 18th Century England', *Centaurus* 8 (1963), 238–62; Shelley Costa, 'The Ladies' Diary: society, gender and mathematics in England, 1704–1754' (Ph.D. dissertation, Cornell University, 2000).

36 *The austere language of mathematics*: Teri Perl, 'The Ladies' Diary or Woman's Almanack, 1704–1841', *Historia Mathematica* 6 (1979), 36–53 at 41, 48; Shelley Costa, 'The "Ladies' Diary": gender, mathematics, and civil society in early-eighteenth-century England', *Osiris* 17 (2002), 49–73 at 49, 71 and *passim*.

36 *problems in* Martin's Magazine: Benjamin Martin (ed.), *Miscellaneous Correspondence* vol. 4 (London, 1764), 767–8 (dated December 1761) and various subsequent appearances up to December 1763. *Public Characters*, vol. 2, p. 103, is the authority for identifying Hutton with Tonthu.

36 *Benjamin Martin*: P.J. Wallis, 'British Philomaths – Mid-eighteenth Century and Earlier', *Centaurus* 17 (1973), 301–14 at 303.

36 *five of his solutions were printed*: The Gentleman's Diary [London, 1763].

36 *home of the hardest*: Pedersen, 'The "Philomath"', *passim*; Joe Albree and Scott H. Brown, '"A Valuable Monument of Mathematical Genius": The Ladies' Diary (1704–1840)', *Historia Mathematica* 36 (2009), 10–47 at 21.

36 The Ladies' Diary: see Costa, 'The Ladies' Diary' (2000).

37 *considerable additions are made*: Charles Hutton (ed.), *Miscellanea Mathematica* (London, 1775), 1.

37 *Having already attended the schools*: Robinson, 'Trends in Education', Appendix 1 (unpaginated), *s.v.* Hutton; Woodhorn, SANT/BEQ/26/1/7/77 (notes of Thomas Wilson, 28 March 1822).

37 *historically motivated programme*: Gregory, 'Memoir', 202.

38 *if a multiplier is itself a product*: Hutton, *Guide*, 28, 30.

39 *But ultimately the aim*: Hutton, *Guide*, 95.

39 *'calculations of the same accounts'*: Hutton, *Guide*, 153.

39 *Hutton paid for the printing himself*: Hutton, *Guide*, title page.

39 *His patron Robert Shaftoe*: Woodhorn, SANT/BEQ/26/1/8/584: Letter of S. Barrass to Thomas Wilson, 1825; Hutton, *Guide*, a2r.

40 *cutting his own type with a penknife*: *Public Characters*, vol. 2, p. 104; 'Biographical Anecdotes of Charles Hutton, L.L.D. F.R.S.', *The Philosophical Magazine* 21 (February 1805), 62–7 at 63.

40 *advertised in a number of newspapers*: *London Evening Post*, 20 March 1764; *London Chronicle*, 12 April 1764; *Public Advertiser*, 16 April 1764.

40 *The Banson dynasty*: Wallis, *Mathematical Tradition*, 16; William Banson, *The Schoolmaster and Scholar's Mutual Assistant* (London and Newcastle, 1760).

40 *Another northern author*: William Emerson, *Cyclomathesis: or an easy introduction to the several branches of the mathematics* (London, 1763).

40 *'been found . . . useful in schools'*: 'Hutton's Arithmetic and Book-Keeping', *General Evening Post*, 23 July 1771.

40 *riding over to the village of Prudhoe*: Woodhorn, SANT/BEQ/26/1/8/584: Letter of S. Barrass to Thomas Wilson, 1825.

41 *Announced in the Newcastle papers*: Proposal for printing the *Mensuration* by subscription, *Newcastle Chronicle*, 26 December 1767.

41 *the list of subscribers*: Hutton, *Mensuration*, v–xiv.

41 *fifteen shillings a book*: Advert for the *Mensuration*, *The Leeds Intelligencer*, 16 April 1771.

41 *'that part of the country'*: Charles Hutton, *The Principles of Bridges* (Newcastle, London and Edinburgh 1772), iv.

42 *'A Line is a length'*: Hutton, *Mensuration*, 1.

42 *'A conoid . . .'*: Hutton, *Mensuration*, 217.

42 *'General Scholium'*: Hutton, *Mensuration*, 28–32.

43 *a note which set out the history*: Hutton, *Mensuration*, 85–9.

43 *typical burst of complexity*: Hutton, *Mensuration*, 234–41.

43 *a letter to one of the Newcastle papers*: letter concerning rules for cutting timber, *Newcastle Courant*, 13 February 1768.

44 *six hundred feet of plaster mouldings*: Hutton, *Mensuration*, 599.

44 *some poor reviews*: Review of the *Mensuration* in *The Critical Review* 32 (October 1771), 286–8.

44 *by far the best treatise*: 'Charles Hutton, LL.D. F.R.S.', in Thomas Leybourn (ed.), *New Series of the Mathematical Repository*, vol. 5 (London, 1830), 187–96 at 189.

44 *O Science!*: Thomas Sadler, 'Science, a poem' (London, 1768), also quoted in P.M. Horsley, *Eighteenth-Century Newcastle* (Newcastle, 1971), 47.

45 *Bewick . . . did some of his first work on the* Mensuration: *Public Characters*, vol. 2, p. 107; 'Memoir of Mr. Thomas Bewick', *The Gentleman's Magazine* (January 1829), 17–20 at 18; Horsley, *Eighteenth-Century Newcastle*, 46 (quote).

45 *102-page book on the theory of bridge building*: Hutton, *Bridges*; see also Hutton, *Plan*; Isaac Farrer, *Narrative of the Great Flood in the Rivers Tyne, Tease, Wear, &c. on the 16th and 17th of Nov. 1771* (Newcastle, 1772).

46 *the reviewers were quick to point out*: Reviews of Hutton, *Bridges* in *The Critical Review* 34 (November 1772), 374–6; *Monthly Review* 48 (January 1773), 71; *The Weekly Magazine* 19 (25 March 1773), 406; *The Town and Country Magazine* 11 (April 1779), 176; *Journal encyclopédique* 35 (February 1773), 534–5.

46 *'We are at a loss to account'*: Review of Hutton, *Bridges* in *The Critical Review* 34 (November 1772), 374–6 at 376.

46 *Older copies of the original* Diary: Charles Hutton, *The Diarian Miscellany* (London, 1775), vol. 1, p. iii.

46 George Coughron: Richard Welford, *Men of Mark 'Twixt Tyne and Tweed* (London, 1895), vol. 1, pp. 634–7; Horsley, *Eighteenth-Century Newcastle*, 29–34.

47 *probably Hutton himself*: 'Biographical Anecdotes', 63.

47 *Samuel Clarke launched a* Diarian Repository: Hutton, *Tracts* (1812), vol. 3, pp. 381–2. An account of Clarke's attacks on various mathematicians in the pages of various periodicals appears in T.T. Wilkinson, 'Notae Mathematicae No. IV', *Mechanic's Magazine* 61 (1854), 243–6.

47 *badly rattled*: Archibald, 'Minor English Mathematical Serials', 382, citing Hutton's annotations on copies of the *Miscellany*.

47 *numbers of his* Miscellany *were delayed*: Advert for no. 3 of the *Miscellany*, *General Evening Post*, 25 February 1772.

47 *Clarke's* Repository *abruptly ceased*: *The Diarian Repository; or, Mathematical Register* (London, 1774), title page indicating coverage of the *Diary* only up to 1760.

48 *the publisher asked one pound nine shillings*: Advert for the *Miscellany* in the *Morning Chronicle and London Advertiser*, 6 March 1776.

48 *heard from a friend*: 'Charles Hutton' in Leybourn, *Mathematical Repository*, 191; Gregory, 'Memoir', 206.

49 *a public competition . . . 'merit alone'*: 'Memoir of the late Dr. Hutton', *The Gentleman's Magazine* (March 1823), 228–32 at 229.

49 *'it is interest alone'*: quoted in Williams, *Life in Georgian England*, 41.

49 *public examinations at the British universities*: Andrew Warwick, *Masters of Theory: Cambridge and the rise of mathematical physics* (Chicago, 2003), 118–22.

49 *'without any interest'*: *Public Characters*, vol. 2, p. 108.

49 *Northumberland was a prominent Tory*: Woodhorn, SANT/BEQ/26/1/8/584: Letter of S. Barrass to Thomas Wilson, 1825; John Cannon, 'Percy [*formerly* Smithson], Hugh, first duke of Northumberland (*bap.* 1712, *d.* 1786), politician' in *ODNB*.

49 *the Earl of Sandwich*: Letter from Isaac Cookson Esq to Mr Thomas Hodgson, printed in Bruce, *Memoir*, 15; N.A.M. Rodger, 'Montagu, John, fourth earl of Sandwich (1718–1792), politician and musical patron' in *ODNB*.

50 *William Emerson*: Bruce, *Memoir*, 15–17.

50 *names he knew*: *Public Characters*, vol. 2, p. 109.

50 *Other candidates*: Gregory, 'Memoir', 206; *Public Characters*, vol. 2, p. 110.

50 *The questions*: Gregory, 'Memoir', 207; *Public Characters*, vol. 2, pp. 110–11.

50 *A persistent rumour*: Woodhorn, SANT/BEQ/26/1/8/584: Letter of S. Barrass to Thomas Wilson, 1825.

51 *particular recommendation*: Gregory, 'Memoir', 207; *Public Characters*, vol. 2, p. 111.

51 *the school in Westgate Street*: Adverts in the *Newcastle Courant*, 5 June and 10 July 1773.

51 *The coach journey*: Moffat and Rosie, *Tyneside*, 210; letters of William Emerson to R. Harrison, 25 Jan 1774 and of Charles Hutton to William Armstrong, 20 Apr 1822, printed (extracts) in Bruce, *Memoir*, 16–17, 42–3.

51 *Almost a million tons of coal*: Fordyce, *A History of Coal*, 105, 108.

4 Professor

52 *Britain's biggest munitions manufactory*: Andrew Saint and Peter Guillery, *Survey of London volume 48: Woolwich* (Yale, 2012), chapter 3 ('The Royal Arsenal'): online draft available at https://www.bartlett.ucl.ac.uk/architecture/research/survey-of-london/woolwich/documents/48.3_The_Royal_Arsenal.pdf, 6–7; O.F.G. Hogg, *The Royal Arsenal: its background, origin, and subsequent history* (London and New York, 1963), vol. 1, p. 493, vol. 2, pp. 1289, 1292.

53 *The Royal Military Academy*: http://www.royal-arsenal-history.com/royal-arsenal-timeline.html; W.D. Jones, *Records of the Royal Military Academy* (Woolwich, 2nd edn, 1895), 4, 8.

53 *fires or unplanned explosions . . .*: Hogg, *The Royal Arsenal* vol. 1, 460–63; Saint and Guillery, 'The Royal Arsenal', 27.

54 *In the warren, or park*: 'Woolwich' (unpaginated) in *The Copper-Plate Magazine* (London, 1792–1802).

54 *His house needed ten pounds' worth of repairs*: Hogg, *The Royal Arsenal*, vol. 1, 380, citing London, Public Record Office, WO/47/81, p. 491 (28 June 1773).

54 *Hutton's family didn't at first accompany him*: See Chapter 5.

55 *a sort of regimental university*: Jones, *Records of the RMA*, 1; Hogg, *The Royal Arsenal*, vol. 1, 345–6; G.A. Shepperd, *Sandhurst: The Royal Military Academy Sandhurst and its Predecessors* (London, 1980), 11.

55 *The chief master, Allen Pollock*: Jones, *Records of the RMA*, 18–20, 22; Hogg, *The Royal Arsenal* vol. 1, 379–80.

55 *Reverend William Green*: Jones, *Records of the RMA*, 22.

56 *predictable problems with their behaviour*: Jones, *Records of the RMA*, 6, 9 (quote), 13, 15, 24, 25.

57 *Window-smashing. 'Pelting'*: 9 (swimming), 15, 26 (hallooing), 13 (windows), 27 (pelting), 6 (middle finger); F.G. Guggisberg, *'The Shop': the story of the Royal Military Academy* (London, 1900), 20 (fun).

58 *The theory classes*: Jones, *Records of the RMA*, 2, 11.

59 *'overcoming by resourcefulness'*: C.A.L. Graham, *The Story of the Royal Regiment of Artillery* (Woolwich, 1928, 4th edn 1983), 22.

59 *Cannon up to ten feet long*: O.F.G. Hogg, *Artillery: its origin, heyday and decline* (London, 1970), 61.

59 *Furious speeds were possible*: B.P. Hughes, *British Smooth-Bore Artillery: the muzzle loading artillery of the 18th and 19th centuries* (London, 1969), 47–9, 78; Graham, *The Story of the Royal Regiment of Artillery*, 14.

60 *a full-scale attack*: Jones, *Records of the RMA*, 2.

60 *General Belford*: Jones, *Records of the RMA*, 26.

60 *'went hand in hand'*: John Gascoigne, *Cambridge in the Age of the Enlightenment* (Cambridge, 1989), 174–5, 297. See also Hans, *New Trends*, *passim*.

61 *'a way to settle in the Mind'*: John Locke, 'Of the Conduct of the Understanding' in *Posthumous Works of Mr. John Locke* (London, 1706), 2–137, at 30 (§7).

61 *the so-called Dissenting Academies*: Irene Parker, *Dissenting Academies in England, their rise and progress and their place among the educational systems of the country* (Cambridge, 1914), *passim*; David A. Reid, *Science and Pedagogy in the Dissenting Academies of Enlightenment Britain* (Ph.D. thesis, University of Wisconsin-Madison, 1999), iii, 1–2, 6.

63 *William Green*: Jones, *Records of the RMA*, 22.

63 *illiterates were not supposed to be admitted*: Jones, *Records of the RMA*, 17.

64 *an entrance exam*: Jones, *Records of the RMA*, 20.

64 *by 1775 his Mensuration was a set text*: Jones, *Records of the RMA*, 23.

65 *Professor Pollock was pensioned off*: Jones, *Records of the RMA*, 24; Hogg, *The Royal Arsenal*, vol. 1, p. 382.

65 *William Green brought to a head*: Jones, *Records of the RMA*, 24–5.

66 *Sixteen companies of the Royal Artillery*: Graham, *The Story of the Royal Regiment of Artillery*, 19.

66 *the shortage of suitable candidates*: Jones, *Records of the RMA*, 22.

66 *cadets were being hurried through*: Jones, *Records of the RMA*, 26 (quote), 28.

67 *on probation to Hutton*: Jones, *Records of the RMA*, 26.

67 *Green was asked to help out*: Jones, *Records of the RMA*, 23, 28, iii (appointment of Bonnycastle as additional mathematical master).

67 *a pay rise*: Jones, *Records of the RMA*, 48.

67 *Discipline wasn't improved*: Jones, *Records of the RMA*, 28, 29.

68 *a rumour went around*: Shepperd, *Sandhurst*, 20–21; Jones, *Records of the RMA*, 27.

68 *Another indirect consequence*: Hogg, *The Royal Arsenal*, vol. 1, p. 451.

68 *Cholick Lane . . . 'Horrid Smells'*: Saint and Guillery, 'The Royal Arsenal', 27; Hogg, *The Royal Arsenal*, vol. 1, p. 463.

69 *chronic headaches*: Guggisberg, *The Shop*, 34; see also Thomas Simpson, ed. Charles Hutton, *Select Exercises for Young Proficients in the Mathematicks* (London, 1792), xxii.

5 Odd-Job Man

70 *I am here almost as recluse*: Letter of Charles Hutton to Robert Harrison, 13 Jan 1779, printed in Melmore, 'Some Letters', 71–4.

70 *Hutton's wife and children*: *Public Characters*, vol. 2, p. 102.

70 *One Newcastle historian was indiscreet*: Mackenzie, *Descriptive and Historical Account*, 560.

71 *some of his friends knew Margaret Ord*: Letter of Charles Hutton to Nevil Maskelyne, 27 Jun 1785: Cambridge University Library, RGO 4/187/11, fos. 1–2.

71 *There were Ords in the Royal Artillery*: Jones, *Records of the RMA*, 74; Southey et. al., *The Ingenious Mr Avison* 24, 59–60; *The History of the British Coal Industry* vol. 2, p. 41; *List of Fellows of the Royal Society 1660–2007*, https://royalsociety.org/~/media/Royal_Society.../fellowship/Fellows1660-2007.pdf, 267.

71 *Hutton's son Harry was in Woolwich*: *List of Officers of the Royal Regiment of Artillery from the Year 1716 to the present date* (Woolwich, 1869), 15 (Henry Hutton, cadet, 2 March 1776; 2nd lieutenant 21 February 1777).

72 *renting a set of rooms in the city*: *An Appeal to the Fellows of the Royal Society* (London, 1784), 12.

73 *a member of the committee*: *Public Characters*, vol. 2, p. 109.

73   *corresponding with Maskelyne's assistant*: Letter of Reuben Burrow to Charles Hutton, 24 Sep 1773: University College London, MS Graves 23/3/5.

73   *'most intimate friends'*: Abraham Rees, *The Cyclopaedia* (London, 1820), vol. 22, s.v. Maskelyne.

73   *Honest and popular*: Mary Croarken, 'Astronomical Labourers: Maskelyne's assistants at the Royal Observatory, Greenwich, 1765–1811', *Notes and Records of the Royal Society of London* 57 (2003), 285–98 at 287; Charles Hutton to Nevil Maskelyne, 27 Jun 1785: Cambridge University Library, RGO 4/187/11, fos. 1–2 (also Charles Hutton to Nevil Maskelyne, 27 Feb 1796: Cambridge University Library, RGO 4/187/26:1); Charles Hutton, *Mathematical Tables* (London, 1785), title page verso and 156, 168–9.

73   The Nautical Almanac: Mary Croarken, 'Tabulating the Heavens: computing the nautical almanac in 18th-century England', *Annals of the History of Computing* (2003); Mary Croarken, 'Providing Longitude for All: the eighteenth century computers of the Nautical Almanac', *Journal of Maritime Research* (October 2002).

73   *The annual books of tables*: Croarken, 'Tabulating the Heavens', 48, 50; Croarken, 'Longitude for All', 114.

74   *a network of human 'computers'*: Croarken, 'Tabulating the Heavens', 54; also Mary Croarken, 'Mary Edwards: computing for a living in 18th-century England', *Annals of the History of Computing* (2003); Thomas Lindsay, 'Historical Sketch of the *Nautical Almanac*' parts V and VI, *Transactions of the Astronomical and Physical Society of Toronto* 9 (1898), 2–10.

75   *a 'comparer' kept an eye on things*: Croarken, 'Tabulating the Heavens', 53, 57, 59; Croarken, 'Longitude for All', 112, 114, 116, 121.

75   *Hutton did all this*: Cambridge University Library, RGO 4/325 54v–55r, account books for the *Nautical Almanac*; Croarken, 'Longitude for All', 117.

76   *to make contact with the Board of Longitude*: Cambridge University Library, RGO 14/5, p. 344 (minutes of 6 March 1779); RGO 14/5, pp. 347–8 (minutes of 10 July 1779); RGO 14/17, pp. 337–9 (account with Charles Hutton, 1779–1782).

76   *and as it seemed probable*: Hutton, *Miscellanea Mathematica*, 341.

77   *The Board of Longitude accepted*: Cambridge University Library, RGO 14/6/2, 6–7 (minutes of 15 July 1780), 16 (3 March 1781), 29 (1 December 1781), 62 (6 March 1784); Croarken, 'Tabulating the Heavens', 55; Charles Hutton, *Tables of the Products and Powers of Numbers* (London, 1781), title page.

77   *The preface displayed*: Hutton, *Tables* (1781), a2r–b1v; review of Hutton, *Tables* (1781) in *Monthly Review* 73 (October 1785), 311–12 at 312.

78   *notorious for its inaccuracy*: Hutton, *Tables* (1785), v–vi, 40, 176, 342–3.

78   *The manuscript of his logarithm tables*: Cambridge University Library, RGO 4/326, with initial calculations and notes on the strategy of calculation

(ff. 4r–33v and 54r–76r) plus fair copies (ff. 34r–35r, 104r–138r, 156v–232v) and extra rough or preliminary work (ff. 102r–v, 141v–154r). The initial calculations on ff. 10r–33v are in a different hand from the bulk of the document. *Public Characters*, vol. 2, p. 120.

79 *for women to work as unpaid assistants*: Croarken, 'Mary Edwards', 14; Museum of the Royal Artillery ('Firepower'), MD 913/5 (manuscript account of Hutton's artillery experiments, 1775, with insertions in another hand); letter of Charles Hutton to Catherine Hutton, 26 Jan 1819: Wellcome Collection, MS 5270 no. 35.

79 *women were visible*: Albree and Brown, 'A Valuable Monument', 17; Perl, 'The Ladies' Diary', 37, 45–6; Wallis, 'Mathematical Tradition', 40; Costa, 'The Ladies' Diary' (2002), 65–71; Croarken, 'Mary Edwards', 9.

79 *Hutton added an enormous preface*: See J.W.L. Glaisher, 'The Earliest Use of the Radix Method for Calculating Logarithms, with historical notices relating to the contributions of Oughtred and others to mathematical notation', *The Quarterly Journal of Pure and Applied Mathematics* 46 (1915), 125–97, at 173, 181–2 for a trace of Hutton reading and annotating in pursuit of this material.

80 *delays in the calculating work*: Hutton, *Tables* (1785), ix, states seven or eight years, but the evidence of the manuscript (Cambridge University Library, RGO 4/326, with dates from 28 September 1784 (fol. 4v) to May 1785 (fol. 76r)) strongly suggests this should be months.

80 *a demanding programme of checking*: Hutton, *Tables* (1785), vi.

80 *a judicious arrangement of the subjects*: Review of Hutton, *Tables* (1785) in *Critical Review* 62 (July and August 1786), 13–17 and 113–17 at 115 (quote); also in *New Review* (1785), 11–15 and *Monthly Review* 74 (May 1786), 344–6 .

81 *some favourable notices*: Notices of Hutton, *Mathematical Miscellany* in *The London Review* 3 (June 1776), 480–81 and *The Town and Country Magazine* 8 (June 1776), 324.

81 *he was approached by the Stationers' Company*: Archibald, 'Minor English Mathematical Serials', 381, 383; *Public Characters*, vol. 2, pp. 113–14; Stationers' Company of London, *The Records of the Stationers' Company 1554–1920* (Cambridge, 1985) Court book M, 483; Court Book O, 77; Series I, Box B, folder 6, items i, iii and iv: drafts dated 22 December 1786, then 1787; unnumbered items in Series I, Box C, folder 4; *A Catalogue of the Curious Mathematical, &c Books of the Late Mr. Edw. Rollinson* [London, 1775], title page; letter of Charles Hutton to Robert Harrison, 19 Mar 1781, printed in Melmore, 'Some Letters', 78–9.

82 *'The enigma on a Candlestick'*: The Ladies' Diary (1775), 31.

82 *reviewed mathematical books*: Hutton, *Tracts, Mathematical and Philosophical* (London 1786), 93; *Public Characters*, vol. 2, p. 114; letter of Charles Hutton

to Robert Harrison, 4 Aug 1779, printed in Melmore, 'Some Letters', 74–6 at 75.

83 *proposed for fellowship of the Royal Society*: London, Royal Society, EC/1774/18.

83 *attended one meeting of the Society out of two*: An Appeal, 12.

83 *Maskelyne had a specific project in hand*: Nevil Maskelyne, 'A Proposal for Measuring the Attraction of Some Hill in this Kingdom by Astronomical Observations', *Philosophical Transactions* 65 (1775), 495–9 at 496 (quote).

84 *Charles Mason . . . Reuben Burrow*: Nevil Maskelyne, 'An Account of Observations Made on the Mountain Schehallien for Finding its Attraction', *Philosophical Transactions* 65 (1775), 500–42 at 502 (quote); John Pringle, *A Discourse on the Attraction of Mountains* (London, 1775), 28.

85 *Maskelyne himself declined*: Gregory, 'Memoir', 211; London, Royal Society, CMO 6, 311 (24 April 1777); 333 (18 June 1778).

85 *immense labour*: Letter of Charles Hutton to Robert Harrison, 4 Aug 1779, printed in Melmore, 'Some Letters', 74–6.

86 *contour lines*: Charles Hutton, 'An Account of the Calculations Made from the Survey and Measures Taken at Schehallien, in Order to Ascertain the Mean Density of the Earth', *Philosophical Transactions* 68 (1778), 689–788 at 756–7; John R. Smallwood, 'Maskelyne's 1774 Schiehallion Experiment Revisited', *Scottish Journal of Geology* 43 (2007), 15–31 at 19.

87 *'weighing the earth' remained*: See chapter 11.

87 *'great quantities of metals'*: Hutton, 'Account of the Calculations', 783.

88 *the Woolwich Military Society*: Museum of the Royal Artillery ('Firepower'), MD 913/5 Item 2., fol. 2r; George Smith, *An Universal Military Dictionary* (London, 1779), s.v. Gunnery; Francis Duncan, *History of the Royal Regiment of Artillery* (3rd edn, London, 1879), 268–70; W. Johnson, 'Charles Hutton, 1737–1823: The prototypical Woolwich Professor of mathematics', *Journal of Mechanical Working Technology* 18 (1989), 195–230 at 224.

88 *Thomas Blomefield*: H.A. Baker, 'Hutton's Experiments at Woolwich, 1783–1791', *Journal of the Arms & Armour Society* 11 (1985), 257–98 at 257, 269, 272, 274–5; Hutton, *Tracts* (1786), 104, 138, 208, 221.

88 *Benjamin Robins*: Benjamin Robins, *New Principles of Gunnery* (London, 1742); Brett D. Steele, 'Muskets and Pendulums: Benjamin Robins, Leonhard Euler, and the ballistics revolution', *Technology and Culture* 35 (1994), 348–82 at 349–54; Hogg, *A History of Artillery*, 46 (quote); H.M. Barkla, 'Benjamin Robins and the Resistance of Air', *Annals of Science* 30 (1973), 107–22 at 107 (quote).

89 *Over the summer of 1775*: Charles Hutton, 'The Force of Fired Gun-Powder, and the Initial Velocities of Cannon Balls, Determined by Experiments', *Philosophical Transactions* 68 (1778), 50–85.

90 *cannon ball breaking to pieces*: Hutton, *Tracts* (1786), 150–51, 193; Hutton, *Tracts* (1812), vol. 3, p. 83.

91 *a first account*: Museum of the Royal Artillery ('Firepower'), MD ,913/5: manuscript account of Hutton's artillery experiments, 1775.

92 *The paper was read to the Royal Society*: Charles Hutton, 'The Force of Fired Gun-Powder', 50.

92 *'those parts of natural philosophy'*: Charles Hutton, 'The Force of Fired Gun-Powder', 52.

92 *'with the more cordial affection'*: John Pringle, *A Discourse on the Theory of Gunnery* (London, 1778), 33.

93 *there was a select number*: *Public Characters*, vol. 2, p. 118; T.E. Allibone, 'The Diaries of John Byrom, M.A., F.R.S., and Their Relation to the Pre-History of the Royal Society Club', *Notes and Records of the Royal Society of London* 20 (1965), 162–83 at 182; Archibald Geikie, *Annals of the Royal Society Club: the record of a London dining-club in the eighteenth & nineteenth centuries* (London, 1917), 143; Charles Hutton, 'Proof of the Failure of the Attempt to Restore Dr. Dodd to life', *The Newcastle Magazine* 1/3 (March 1822), 127–8 at 127 (quote).

6 Foreign Secretary

94 *8 January 1784*: Paul Henry Maty, *An Authentic Narrative of the Dissensions and Debates in the Royal Society* (London, 1784), 66; London, Royal Society, MM/1/46a; *Supplement to the Appeal to the Fellows of the Royal Society* [London, 1784], 14; Andrew Kippis, *Observations on the Late Contests in the Royal Society* (London, 1784), 82. On the 'Dissensions' see also Benjamin Wardhaugh, 'Charles Hutton and the "Dissensions" of 1783–84: scientific networking and its failures', *Notes and Records of the Royal Society* 71 (2017), 41–59.

95 *an upset about the secretaryship*: *An Appeal*, 4–5; *Supplement*, 12; Kippis, *Observations*, 27–8; [John Strange], *Canons of Criticism, extracted from the beauties of Maty's Review* (London, [1784]), 73; Royal Society CMO/7, pp. 12–15 (10 Dec 1778, 14 Jan 1779); Thomas Seccombe, rev. Rebecca Mills, 'Maty, Paul Henry (1744–1787), librarian' in *ODNB*; letter of Charles Hutton to Robert Harrison, 13 Jan 1779, printed in Melmore, 'Some Letters', 71–4.

95 *foreign secretary occupied a back-room position*: *An Appeal*, 6–8; Kippis, *Observations*, 30; *Supplement*, 7; Strange, *Canons*, 69.

96 *a member of the Society's Council*: Royal Society, CMO/7, 15–60 *passim*.

96 *an honorary doctorate*: *A Catalogue of the Graduates in the Faculties of Arts, Divinity, and Law, of the University of Edinburgh since its foundation* (Edinburgh, 1858), 257; letter of Charles Hutton to Harrison, 4 Aug 1779, printed in Melmore, 'Some Letters', 74; Bruce, *Memoir*, 17.

97 *denied that there had been any specific incident*: An Appeal, 23.

98 *'with freedom and firmness'*: An Appeal, 28; Royal Society, MM/1/46a.

98 *He retained a northern accent*: Obituary of Charles Hutton in the *Literary Gazette* (1 February 1823), 75.

98 *One journalist made a joke*: Notice of Hutton, *Principles of Bridges* in *The Town and Country Magazine* 11 (April 1779), 176; 'Sir Joseph Banks', *New Times* (14 July 1820), 4.

98 *mathematics never lent itself*: John Barrow, *Sketches of the Royal Society, and Royal Society Club* (London, 1849), 4–5; Warwick, *Masters of Theory*, 36; Royal Society, JBO, vol. 28, pp. 448, 489; vol. 29, pp. 225, 228, 489 (mathematical papers handled in various ways).

99 *'waspish and petulant expressions'*: [Olinthus Gregory], 'A Review of some Leading Points in the Official Character and Proceedings of the late President of the Royal Society', *Philosophical Magazine* series 1, no. 56 (1820), 161–74, 241–57, at 166.

99 *'done little but apply Conic Sections'*: National Library of Wales, MS 12415C (letter of Joseph Banks to John Lloyd, 23 Feb 1780), quoted in David Philip Miller, '"Into the Valley of Darkness": reflections on the Royal Society in the eighteenth century', *History of Science* 27 (1989), 155–66 at 166.

99 *Banks wanted to consolidate*: David Philip Miller, 'Between Hostile Camps: Sir Humphry Davy's presidency of the Royal Society of London, 1820–1827', *The British Journal for the History of Science* 16 (1983), 1–47 at 4–5; David Philip Miller, 'Sir Joseph Banks: an historiographical perspective', *History of Science* 19 (1981), 284–92, at 288, 291; Miller, 'Into the Valley of Darkness', 162–3.

99 *a dispute about the best shape*: [Gregory], 'A Review', 170–2; William Temple Franklin, *Memoirs of the Life and Writings of Benjamin Franklin* (London, 1818), vol. 1, pp. 322–4; A. Hunter Dupree, *Sir Joseph Banks and the Origins of Science Policy* (Minneapolis, 1984), 15; Anonymous, *An History of the Instances of Exclusion from the [R]oyal Society* (London, 1784), 24; 'Peter Pindar' [John Wolcot], *Peter's Prophecy* (Dublin, 1789), 19; Simon Snip, *The Philosophical Puppet Show* [London, 1783?], 7–8.

100 *American liberty*: letter of Charles Hutton to Robert Harrison, 13 Jan 1779, printed in Melmore, 'Some Letters', 71–4.

100 *Hutton should not be styled 'professor'*: Royal Society CMO/7, 81 (2 August 1781), 59 (16 Nov 1780), 89 (15 Nov 1781); Maty, *Narrative*, 108–9.

100 *foreign correspondence was not being dealt with*: Royal Society, CMO/7, 97–8 (24 Jan 1782), 101–2, 105–6 (21 March 1782), 109 (25 April 1782); An Appeal, 10–11; Kippis, *Observations*, 33.

101 *excursions in international* politesse: An Appeal, 12; Maty, *Narrative*, 115; Kippis, *Observations*, 63.

101 *polite he could do*: Charles Hutton to Kleinschmidt, 13 May 1784: London, Royal Society, RSL/2, no. 27, fol. 2v.

101 *he complied stolidly*: George Herbiniaux, *Traité sur divers accouchemens laborieux* (Brussels, 1782), vol. 2, 195–6 has a specimen of an acknowledgement as sent and signed by Hutton, dated 20 June 1782.

101 *More complex letters*: Kippis, *Observations*, 44–5, 51; Maty, *Narrative*, 102.

102 *something of a stand-off*: Kippis, *Observations*, 65, 67–8; *An Appeal*, 23.

102 *Joseph Banks a problematic figure*: Instances of Exclusion, 4–7 and *passim*; Kippis, *Observations*, 90–99; [Gregory], 'A Review', 174.

102 *Banks blocked the election*: Instances of Exclusion, 5–6; Maty, *Narrative*, 56–9; 'Peter Pindar', *Peter's Prophecy*, A2r.

103 *other kinds of conduct*: Instances of Exclusion, 19–24.

103 *the structure of the Royal Society*: Neil Chambers, *Scientific Correspondence of Sir Joseph Banks, 1795–1820* (London, 2007), vol. 1, p. xxiv.

104 *told Council there was a problem*: *An Appeal*, 12–13; Kippis, *Observations*, 35; *Supplement*, 3, 6, 13; Royal Society MM/1/29, MM/1/42 and CMO/7 150 (minutes of 20 November 1783).

104 *Understanding, Sir, that the circumstance*: *Appeal*, 14.

104 *The house slate was voted in*: Kippis, *Observations*, 8–9, 36–7, 110, 112; Royal Society JBO 31, 260–2; *An Appeal*, 14–15; *Supplement*, 3; *An History*, 17–18; Maty, *Narrative*, 6.

105 *Edward Poore . . . moved thanks*: Maty, *Narrative*, 7–9, 152–3; *An Appeal*, 15–18, 24; *Supplement*, 5; Kippis, *Observations*, 9, 38–40; Royal Society MM/1/40, MM/1/48e, MM/1/42, JBO 31, 265. On the offended foreigner, Bonnet: Maty, *Narrative*, 89, 101–2; Strange, *Canons* 29–39; Sir Gavin de Beer and R.M. Turton, 'John Turton, FRS, 1735–1806', *Notes and Records of the Royal Society of London* 12 (1956–7), 77–97, at 86–8.

105 *a written defence of his conduct*: *An Appeal*, 19–20, 23–4; Maty, *Narrative*, 10, 18, 24, 153; Kippis, *Observations*, 10–11, 42–3, 77; *Supplement*, 7; Royal Society MM/1/40, CMO/7 152, JBO 31, 267, 268–9.

105 *Horsley took the floor again*: Maty, *Narrative*, 12, 20–21; *An History*, 21; Kippis, *Observations*, 114–15; *An Appeal*, 24; Royal Society MM/1/47d; Henry Craik, *English Prose* (New York and London, 1893–6), s.v. Horsley; Anon, 'Biographical Sketch of the life of Sir Joseph Banks . . . ', *Agricultural Magazine* 9 (1811), 333–41 at 339.

106 *Banks's friends assembled*: Chambers, *Scientific Correspondence*, vol. 2, pp. 236–46 (letters 450–58 between Blagden and Banks, 22–30 Dec 1783), 249–50 (letters 461–2 of Kippis and Henry Stebbing to Banks, 31 Dec 1783); Warren Dawson (ed.), *The Banks Letters: A calendar of the manuscript correspondence of Sir Joseph Banks preserved in the British Museum, British Museum (Natural History) and other collections in Great Britain* (London, 1958), 533 (John Coakley Lettsom to Banks, 30 Dec 1783); Kippis, *Observations*, 80; *An Appeal*, 25; Royal Society MM/1/29, MM/1/46, MM/1/47; Firepower MS 913/3 item 2 (quote).

106 *a series of letters*: collected in *Supplement to the Appeal*.

107 *The house was visibly packed*: Maty, *Narrative*, 22, 24–7, 31–3, 37, 40–43, 45, 61–4, 67, 69, 72–3, 76–7; *Supplement*, 9–11, 14; Kippis, *Observations*, 13, 82–3; *An History* 26; Royal Society JBO 31, 269–71; [Gregory], 'A Review', 245; Chambers, *Scientific Correspondence* vol. 2, p. 243.

108 *On 29 January a sly motion*: *Authentic*, 80; *Appeal Supplement*, 14; Kippis, *Observations*, 85; Royal Society, MM/1/46a, MM/1/41, fol. 1v.

108 *On 12 February Francis Maseres . . .*: *Authentic*, 79, 85–134; Kippis, *Observations*, 18, 39, 49, 86, 145–6; *Appeal*, 26; Royal Society, MM/1/30 (quote f. 1r), MM/1/34, MM/1/41, fol. 2r.

109 *on 26 February*: *Authentic*, 134–49; Kippis, *Observations*, 20–21; Royal Society, MM/1/31 and 31a, MM/1/41, fol. 2r, MM/1/46a, b, d.

109 *The manner which he assumed*: Kippis, *Observations*, 145.

109 *'howsoever respectable mathematicks'*: Royal Society, MM/1/46a.

110 *Tis Horsley's voice loud strikes the ear*: The Royal Artillery Museum ('Firepower'), MD/913/3c.

111 *a bizarre incident on 25 March*: *An History*, 27; Royal Society MM/1/47c and JBO 371–2 (with a very different account of the incident). See also Chambers, *Scientific Correspondence*, vol. 2, pp. 263–4 (letter 473, Blagden to Banks, 1 Mar 1784) and 266–7 (letter 476, Banks to Blagden 6 Mar 1784); *An History*, 9; and Royal Society MM/1/32, referring to Maty's behaviour at the meeting of 4 March.

111 *Hutton then offered himself*: Chambers, *Scientific Correspondence*, vol. 2, p. 272, letter 483; *Supplement* 12; Strange, *Canons* 65–70; Kippis, *Observations*, 21–2; Royal Society MM/1/33, MM/1/48, JBO 378–9, 409–10, CMO/7 161.

112 *Hutton himself wrote*: Chambers, *Scientific Correspondence*, vol. 3, pp. 296–7.

112 *literally all over the world*: Chambers, *Scientific Correspondence*, vol. 2, pp. 200–02 (letter 424, Blagden, 30 Oct 1783), 281–2 (letter 491, John Hope to Banks, Edinburgh, 13 May 1784), 283–4 (letter 493, Abbé Theodore Augustin Mann, Brussels, 4 June 1784), 309–10 (letter 514, Banks to Jacques-Pierre Brissot de Warville, 23 Sep 1784); Dawson, *Banks Letters* 154 (Jacques-Pierre Brissot de Warville to Banks, Newman Street, 20 Apr 1784), 158 (Pierre Marie Auguste Broussonet to Banks, Paris, 11 Feb 1784), 184 (Thomas Bugge to Banks, Copenhagen, 12 Oct 1784), 523 (Louis Léon Félicité, Comte de Lauraguais, Duc de Brancas to Banks, Brussels, 4 Apr 1785), 724 (Patrick Russell to Banks, Vizagapatam, 26 Dec 1784, 9 July and 4 Aug 1785).

112 *the Copley Medal for 1784*: Gregory, 'A Review', 167–9.

7 Reconstruction

114 *grazing, turf and wood*: Saint and Guillery, *Survey of London volume 48: Woolwich*, chapter 10 'Woolwich Common and Royal Military Academy

Areas', online draft available at https://www.bartlett.ucl.ac.uk/architecture/ research/survey-of-london/woolwich/documents/48.10_Woolwich_Common_ and_Royal_Military_Academy_Areas.pdf, 1–2.

115 *a 'learned empire'*: David Philip Miller, 'The Royal Society of London 1800– 1835: a study in the cultural politics of scientific organization' (Ph.D. thesis, University of Pennsylvania, 1981), 9; David Philip Miller, 'The Revival of the Physical Sciences in Britain, 1815–1840', *Osiris* 2 (1986), 107–34 at 109.

115 *George's minister for science*: John Gascoigne, 'Banks, Sir Joseph, baronet (1743–1820), naturalist and patron of science' in *ODNB*.

116 *Neither he nor anyone else*: lists of the Royal Society were printed annually with the *Philosophical Transactions* during this period. Contemporary sources such as [Gregory], 'A Review' hint at an organised withdrawal from meetings, but London, Royal Astronomical Society, William Herschel Correspondence 13.H.38 (letter from Hutton to William Herschel, 31 Jan 1785), printed in Wardhaugh, *Correspondence*, 44, seems to indicate otherwise.

116 *a club of their own*: Cambridge University Library, RGO 4/187/11:1–2 (letter of Hutton to Maskelyne, 27 June 1785); MS Add. 7886/117 (letter of Hutton to William Frend, 21 May 1791); RGO 4/187/18: (letter of Hutton to Maskelyne, 19 Dec 1793); RGO 4/187/22: 22:1–2 (letter of Mr Rowed of the Globe Tavern, Fleet Street to Maskelyne, Jun 1795); British Library, Add. MS 37915, ff. 218–19 (letter of Hutton to Windham, 15 Sep 1802); see also Derek Howse, *Nevil Maskelyne: the seaman's astronomer* (Cambridge, 1989), 161.

116 *another 'rebellious dining club'*: Harold B. Carter, *Sir Joseph Banks, 1743– 1820* (London, 1988), 133, 146, 574.

116 *Banks's opponents*: Wardhaugh, 'Charles Hutton and the "Dissensions"'.

117 *Paul Henry Maty died*: Seccombe, 'Maty, Paul Henry'.

117 *Horsley achieved preferment*: Robert Hole, 'Horsley, Samuel (1733–1806), bishop of St Asaph' in *ODNB*.

117 *flattering reprints*: Francis Maseres, *Scriptores Logarithmici* (6 vols: London, 1791–1807), vol. 3, pp. 165–8 (a piece on the area of the circle, from Hutton's *Mensuration*), 207–16 (one of Hutton's *Transactions* papers, on series); also vol. 6, pp. 451–74 ('A Computation of the Length of the Sine of a Circular Arc of One Minute of a Degree', done by Hutton in 1777 but not previously published).

117 *Banks visited Hutton at Woolwich*: Letter of Joseph Banks to Charles Hutton, ?1784: Wellcome Collection MS 5270 no. 5.

117 *Banks was invited to attend*: Letter of Charles Lennox, Duke of Richmond and Lennox to Joseph Banks, 31 May 1786, in Warren R. Dawson, *Supplementary Letters of Sir Joseph Banks* (London, 1962), 59.

117 *Banks and Hutton even managed*: Letter of Charles Hutton to Joseph Banks, 12 Mar 1797, British Library, Add. MS 8098, fol. 385.

118 *the Royal Society of Edinburgh*: C.D. Waterston and A. Macmillan Shearer, *Former Fellows of the Royal Society of Edinburgh, 1783–2002* (Edinburgh, 2006) (https://www.royalsoced.org.uk/cms/files/fellows/biographical_index/fells_indexp1.pdf), vol. 1, p. 469; 'Naamlyst van de tegenwoordige heeren Leden, naar orde van het inkomen', *Verhandelingen, uitgegeeven door de Hollandsche Maatschappye der Weetenschappen, te Haarlem* 29 (1793); *American Academy of Arts and Sciences: Academy Members 1780–present* (http://www.amacad.org/publications/BookofMembers/ChapterH.pdf), 282; Arthur Donovan and Joseph Prentiss, 'James Hutton's Medical Dissertation', *Transactions of the American Philosophical Society* 70 (1980), 3–57 at 7; T.H. Levere and G. L'E. Turner, *Discussing Chemistry and Steam: the minutes of a coffeehouse philosophical society* (Oxford, 2002), 57 and *passim*.

118 *Reuben Burrow*: T.T. Wilkinson, 'Mathematics and Mathematicians, the journals of the late Reuben Burrow', *London, Edinburgh, and Dublin Philosophical Magazine*, 4th ser., vol. 5 (1853), 185–93, 514–22 and vol. 6 (1853), 196–204; Archibald, 'Minor English Mathematical Serials', 389.

119 *He accused Hutton of stealing*: Reuben Burrow, *A Companion to the Ladies Diary* (London, 1781), 32; Reuben Burrow, *The Lady's and Gentleman's Diary* (1776) (quote in preface); Wilkinson, 'Mathematics and Mathematicians', 187 (quote); Howse, *Maskelyne*, 140 (quote from De Morgan).

119 *William Blake, in an early satire*: William Blake, 'The Island in the Moon', online at blakearchive.org, fos. 5, 2, 11, 12.

120 *Hutton and his friends toyed*: Letter of Charles Blagden to Joseph Banks, 31 Oct 1784, in Dawson, *Banks Letters*, 65; Letter of Charles Blagden to Joseph Banks, 6 Nov 1784, in Chambers, *Scientific Correspondence*, vol. 2, pp. 328–9.

120 *soliciting and collecting material*: Royal Artillery Museum MS 913/5f (unattributed press clipping of 1784 or 1785); Chambers, *Scientific Correspondence*, vol. 3, pp. 92–4, 105–6, 110–11 (letters 605, 612, 614, Blagden to Banks, 30 Sep, 23 Oct, 30 Oct 1785).

121 *at his own expense*: Supplement to the Ladies Diary (1788), 32.

122 *Hutton did not keep up*: Supplement to the Ladies Diary (1792), 27.

122 *Even the elusive Margaret Ord*: Supplement to the Ladies Diary (1791), 108–9 and (1794), 15.

122 *Hutton was left with just a handful*: Catalogue of the . . . Library of Charles Hutton.

122 *suggestions for its management*: Letter of Lewis Evans to Charles Hutton, 28 April 1788: draft in Oxford, Museum of the History of Science, MS Evans 31; *Ladies' Diary* (1789), 31; *Ladies' Diary* (1790), 46; *Supplement to the Ladies Diary* (1789), 25.

123 *could easily have appeared*: Review of Hutton, *Tracts* (1786) in *The Critical Review* 63 (January 1787), 1–6 at 2.

123 *kind words in a couple*: Review of Hutton, *Tracts* (1786) in *The Critical Review* 63 (January 1787), 1–6 at 1, 6 (quotes); also in the *Monthly Review* 76 (June 1787), 484–9.

124 *papers to the Royal Society of Edinburgh*: Charles Hutton, 'Experiments on the Expansive Force of Freezing Water, made by Major Edward Williams of the Royal Artillery, at Quebec in Canada . . . . Communicated in a letter from Charles Hutton . . . to Professor John Robison, General Secretary of the Royal Society of Edinburgh', *Transactions of the Royal Society of Edinburgh* 2 (1790), 23–8 (reprinted in *The Literary Magazine and British Review* 6 (January 1791), 20–2); Charles Hutton, 'Abstract of Experiments Made to Determine the True Resistance of the Air to the Surfaces of Bodies, of various figures, and moved through it with different degrees of velocity', *Transactions of the Royal Society of Edinburgh* 2 (1790), 29–36; letter of John Playfair to Charles Hutton, 21 Apr 1788 in Wellcome Collection MS 7430 no. 38.

124 *exchanging manuscripts with friends*: Trinity College Cambridge, MS R.1.59: 87r–105v (copy of 'Mr Maseres's Problem of a Vibrating Line' together with Hutton's own notes and related matter from Henry Cavendish); 155r–165r (copy of Henry Cavendish, 'On Finding the Orbits of Comets in a Parabola from 3 Observations', together with associated correspondence).

124 *an observation at the Royal Observatory*: *Astronomical Observations, made at the Royal Observatory . . . from* MDCCLXXV *to* MDCCLXXXVI (London, 1787), vol. 2, p. 120.

124 *reading furiously*: 'Biographical Anecdotes', 64.

124 *cousin of the nonconformist leader*: obituary of Charles Hutton in the *Literary Gazette* (1 February 1823), 75–6 at 75 (with an obvious mistake giving 'Charles' for 'James'; the reference to Cosway's portrait of James Hutton holding his ear trumpet makes it clear who is intended).

124 *the popular novelist Catherine Hutton*: Charles Hutton to Catherine Hutton, 26 Jan 1819 (Wellcome Collection, MS 5270 no. 35); Charles Hutton to Catherine Hutton, 18 Oct 1819 (Birmingham Archives, Heritage and Photography Service MS 3597/73); Catherine Hutton, *Reminiscences of a Gentlewoman of the Last Century* (Birmingham, 1891), 179–81; note on the two surviving Hutton daughters, Isabella and Catherine, in *The Monthly Magazine* 26 (August 1838), 199–210; Llewellynn Jewitt, 'Pedigree of the family of Hutton, of Derby, Birmingham, etc., etc.', *The Reliquary* (1871), foldout before 215.

125 *Lord Stanhope left Hutton*: Public Record Office, PROB 11/1590/64: Will of The Right Honorable Charles Earl Stanhope of Doctors Common, Middlesex (dated 22 November 1805 and proved 5 March 1817); obituary of Charles Hutton in the *Literary Gazette* (1 February 1823), 75–6 at 75.

125 *cousin James Hutton's grandmother*: London, University College archives,

MS Galton 2/4/1/2/9 (letter of Charles Blacker Vignoles to Francis Galton, 17 Nov 1865); Royal Society MC/3/150 (letter of Charles Blacker Vignoles to the Marquis of Northampton, 25 Mar 1841; copy in Portsmouth History Centre, Vignoles papers, letter 751).

125 *the tale had the Hutton family*: Gregory, 'Memoir', 201.

125 *Hutton's wife Isabella died*: Sykes, *Local Records*, 336, giving the date as 26 May; *Newcastle Magazine* (May 1785), 240; Mackenzie, *Historical Account of Newcastle*, 560.

125 *She became Margaret Hutton*: England, *Births and Christenings, 1538–1975*, and *Crisp's Index* (licence dated 22 July); also *London and Surrey, England, Marriage Bonds and Allegations, 1597–1921*, accessed via ancestry.co.uk.

126 *I was the last to consent*: C.F. Adams (ed.), *The works of John Adams, second president of the United States* (Boston, 1850–6), vol. 8, pp. 255–7, quoted in John Cannon, 'George III (1738–1820), king of the United Kingdom of Great Britain and Ireland, and king of Hanover' in *ODNB*.

126 *Servicing public debt*: John Brewer, *The Sinews of Power: war, money and the English state, 1688–1783* (London, 1989), 94.

126 *the British Army was sharply reduced*: Graham, *The Story of the Royal Regiment of Artillery*, 21.

127 *no commissions in sight*: Shepperd, *Sandhurst*, 21.

127 *three 'academies'*: Guggisberg, 'The Shop', 31.

127 *Hutton wasn't one to stick rigidly*: Charles Hutton, *A Course of Mathematics* (London, 1798), vol. 1, p. iv.

127 *Howard Douglas*: Fullom, *General Sir Howard Douglas*, 16–17.

128 *As a preceptor*: Gregory, 'Memoir', 219.

128 *the unhealthiness of the Woolwich site*: Simpson, *Select Exercises*, xxii.

128 *leave to move further away*: Hogg, *The Royal Arsenal*, vol. 1, p. 388, citing Public Record Office WO/47/109, p. 860; Gregory, 'Memoir', 218.

129 *a few other officers' houses*: Saint and Guillery, 'Woolwich Common', 2, 5, 10.

129 *Hutton had little hesitation*: Bruce, *Memoir*, 20; Saint and Guillery, 'Woolwich Common', 5; Charles Hutton, 'Note on the Divining Rod', *Philosophical Magazine* 55/266 (June 1820), 466.

129 *Even the bricks and slates*: *Public Characters*, vol. 2, pp. 122–3.

129 *Maskelyne lent him*: Cambridge University Library, REG 9/37: 2 (Notebook of accounts of Nevil Maskelyne), 83.

130 *the Cube House*: Saint and Guillery, 'Woolwich Common', 6, but the statement there that Hutton did not live in the Cube House is contradicted elsewhere: Hutton, 'Note on the Divining Rod', 466; letter of Charles and Isabella Hutton to Charles Blacker Vignoles, 22 Aug and 28 Sep 1814, in Portsmouth History Centre, Vignoles Papers.

130 *made his fortune*: Bruce, *Memoir*, 22.

8 A Military Man

131 *Woolwich Common, 13 September 1787*: Hutton, *Tracts* (1812), vol. 3, p. 84.

131 *'Our British university'*: 'Woolwich' (unpaginated) in *The Copper-Plate Magazine* (London, 1792–1802); also obituary of Charles Hutton in *The Edinburgh Annual Register* 16 (December 1823), 328–31 at 329.

132 *Private academies adopted its textbooks*: An Abstract of the Course of Education, taught at the Royal Military and Marine Academy, at Belmont on Summer-Hill, Dublin (Dublin, 1784).

132 *The foundation of the Royal Military College*: Shepperd, *Sandhurst*, 11, 21, 24, 28.

132 *the East India Company's training college*: H.M. Vibart, *Addiscombe, its heroes and men of note* (Westminster, 1894), 5; Jones, *Records of the RMA*, 47–8, 63.

133 *'Brief, yet comprehensive'*: Charles Hutton, *The Compendious Measurer* (London, 1786), title page, iii.

133 *'my plumber'*: Hutton, *Compendious Measurer*, 311, 307.

134 *a* Key . . . *to his* Guide: Charles Hutton, *A Key to Hutton's Arithmetic* (London, 1786).

134 *as far as possible, reused the same wording*: Charles Hutton, *Elements of Conic Sections* (London, 1787), ix.

134 *the odd mistake*: Hutton, *Conic Sections*, 69 (text wrongly reprinted from the 'ellipse' section under corollary 3).

134 *Repeat the working*: Hutton, *Conic Sections*, 161, 135, 152.

134 *It circulated as a manuscript*: Hutton, *Conic Sections*, vii.

135 *'I am much pleased'*: Letter of John Playfair to Charles Hutton, 21 Apr 1788, Wellcome Collection MS 7430 no. 38.

135 *presented at court*: 'Biographical anecdotes', 67; *Public Characters*, vol. 2, p. 121.

136 *gunnery teaching*: Such as Oxford, Bodleian MS Eng. Misc. e 146, treatise 'On the Mathematical Principles of Gunnery' by William Lambton, c. 1781–2; Reuben Burrow, *A Restitution of the Geometrical Treatise of Apollonius Pergæus on Inclinations, also the theory of gunnery* (London, 1779), xxv.

136 *the parabola theory therefore provided*: Hogg, *A History of Artillery*, 41; Hutton, *Course*, vol. 2, p. 163; Hutton, *Tracts* (1812), vol 3, p. 266.

136 *Benjamin Robins*: Benjamin Robins, *New Principles of Gunnery* (London, 1742; ed. Hutton, 1805); Steele, 'Muskets and Pendulums'.

136 *how muzzle speed depended . . .* : Hutton, *Tracts* (1786), 249–62.

137 *a machine to test and quantify*: Hutton, *Tracts* (1812), 153–63; Seymour H. Mauskopf, 'Chemistry in the Arsenal: state regulation and scientific methodology of gunpowder in eighteenth-century England and France', in *The Heirs*

of Archimedes: science and the art of war through the Age of Enlightenment, ed. Brett D. Steele and Tamera Dorland (Cambridge, MA, 2005), 293–330 at 312; Baker, 'Hutton's Experiments at Woolwich, 1783–1791', 257–98 at 259, 290.

137 *firing balls down the Thames*: Hutton, *Tracts* (1786), 216–20 and *passim*.

137 *the shot were wandering*: Hutton, *Tracts* (1786), 223–4.

137 *the quality of 'windage'*: Hutton, *Tracts* (1812), vol. 3, p. 261; also Douglas, *Naval Gunnery*, 154.

138 *the Magnus effect*: W. Johnson, 'Benjamin Robins: a neglected mid-18th century military engineer-scientist', in *Collected Works on Benjamin Robins and Charles Hutton* (New Delhi, 2001), 1–12 at 8; W. Johnson, 'The Magnus Effect: early investigations and a question of priority', in *Collected Works*, 13–32, *passim*.

138 *a cannon that was itself*: Hutton, *Tracts* (1786), 107–8 and *passim*.

138 *Robins's old apparatus*: Hutton, *Tracts* (1812), vol. 3, 163–208.

139 *Hutton received detailed reports*: 'Firepower', MS 913/2.

139 *the results were disappointing*: Hutton, *Tracts* (1812), vol. 3, pp. 209–315.

139 *the two-and-one-tenth power*: Hutton, *Tracts* (1812), vol. 3, p. 225.

140 *He drew a picture*: Hutton, *Tracts* (1812), vol. 3, pp. 267–9, 274–8.

141 *exercises that he included*: Hutton, *Course*, vol. 2, p. 164.

141 *'Military tactics have been much benefited'*: Obituary of Hutton in *London Magazine* 7 (March 1823), 368; also obituary of Hutton in *The Literary Chronicle* 5 (1 February 1823), 77; Douglas, *Naval Gunnery*, 27, 32, 139.

141 *short-nosed 'carronades'*: Hutton, *Tracts* (1786), 103; Robins, *New Principles* (1805), 36–7; Charles Hutton, George Shaw and Richard Pearson (eds), *The Philosophical Transactions . . . abridged* (London, 1809), vol. 18, p. 172; review of Hutton, *Tracts* (1812) in *The Quarterly Review* (1813), 400–18 at 414.

141 *his opinion about reducing windage*: Douglas, *Naval Gunnery*, 92; essay review including Hutton, *Tracts* (1812), in *The British Review* 20 (1822), 283–300 at 289–90.

141 *Henry Shrapnel and William Congreve*: Hogg, *A History of Artillery*, 51; Steele, 'Military "Progress"', 372; Hughes, *British Smooth-Bore Artillery*, 56; *The Story of the Royal Regiment of Artillery*, 14; Simon Werrett, 'Congreve's Rational Rockets', *Notes and Records of the Royal Society* 63 (2009), 35–56, *passim*.

141 *an important site*: Saul David, *All the King's Men: the British redcoat in the era of sword and musket* (London, 2012), 368–9; Johnson, 'Prototypical', 218–19; Mauskopf, 'Chemistry in the Arsenal', 308, 311; Werrett, 'Rational Rockets', 37, 39; Niccolò Guicciardini, *The Development of Newtonian Calculus in Britain 1700–1800* (Cambridge, 1989), 109; Jenny Bulstrode, 'The Promiscuous exercises of the Woolwich Bois Boys', paper given at All Souls College, Oxford, December 2015.

142 *actual landings*: Catriona Kennedy, *Narratives of War: military and civilian experience in Britain and Ireland, 1793–1815* (Basingstoke, 2013), 166 and *passim*.

142 *forge a nation*: Linda Colley, *Britons: forging the nation* (London, 2003); Kennedy, *Narratives of War*.

142 *nine years of economy*: David, *All the King's Men*, 313.

142 *a whole breed*: C.J. Esdaile, 'The British Army in the Napoleonic Wars: approaches old and new' [essay review], *English Historical Review* 130 (2015), 123–137 at 125.

142 *one man in four*: Kennedy, *Narratives of War*, 5.

142 *Professional families flooded*: Mark S. Thompson, 'The Rise of the Scientific Soldier as Seen Through the Performance of the Corps of Royal Engineers During the Early 19th Century' (Doctoral thesis, University of Sunderland, 2009), 117; Esdaile, 'The British Army', 124.

143 *the award in 1833*: W.Y. Carman and Michael Roffe, *The Royal Artillery* (Reading, 1973), 14.

143 *William Congreve*: Carman and Roffe, *The Royal Artillery*, 11.

143 *ten thousand rounds in a day*: *The Story of the Royal Regiment of Artillery*, 34.

143 *A new horse artillery . . . new battalions*: Thompson, 'The Scientific Soldier', 91; Carman and Roffe, *The Royal Artillery*, 11.

143 *the Royal Engineers*: Thompson, 'The Scientific Soldier', *passim*.

143 *The number of cadets increased*: Jones, *Records of the RMA*, 56, 57.

143 *complaints . . . about overcrowding*: Jones, *Records of the RMA*, 43.

143 *'so small as to be insufficient'*: Charles Hutton, *A Mathematical and Philosophical Dictionary containing an explanation of the terms, and an account of the several subjects, comprized under the heads mathematics, astronomy, and philosophy both natural and experimental* (2 vols: London 1795–1796), vol. 1, p. 17; Jones, *Records of the RMA*, 43, 61.

144 *to ask rather desperately*: Jones, *Records of the RMA*, 44.

144 *the cadets of the East India Company*: Jones, *Records of the RMA*, 47–8.

144 *the Irish corps of artillery*: Jones, *Records of the RMA*, 50–51.

144 *to lecture on natural philosophy*: Jones, *Records of the RMA*, 49.

144 *the summer vacation*: Jones, *Records of the RMA*, 54; Thompson, 'The Scientific Soldier', 49.

144 *'Disgraceful irregularities'*: Jones, *Records of the RMA*, 43–4, 34, 41; also William Saint, *Four Letters to Lieutenant Colonel Mudge* (London, 1811), 28.

145 *two new mathematical assistants*: Jones, *Records of the RMA*, 53.

145 *management of students' progression*: Jones, *Records of the RMA*, 56, 61; Saint, *Four Letters*, 13–14.

145 *real utility to the war effort*: Thompson, 'The Scientific Soldier', 51, 77, 81.

146 *his daughters Charlotte and Eleanor*: Obituary of Charlotte Hutton in *The Gentleman's Magazine* (October 1794), 960–61; letter of Eleanor Wills (née Hutton) to Charles Blacker Vignoles, December 1816, in Portsmouth History Centre, Vignoles Papers, Letter 127 (this is the only evidence for Eleanor's residence in France). Charlotte was baptised (aged 12) on 4 September 1790, at St Clement Danes Westminster: parish register accessed via www.find-mypast.co.uk.

146 *Charlotte was a particularly welcome addition*: Obituaries of Charlotte Hutton in *The Gentleman's Magazine* (October 1794), 960–61; in the *Hibernian Magazine* (1794), 477–8; 'Hutton, Charlotte' in W.M. Johnson and Thomas Exley, *The New Imperial Encyclopaedia* (London, n.d.), vol. 3, pp. 60–61.

146 *Charles Henry Vignoles*: Letter of Charles Henry Vignoles to Charles Hutton, 31 October 1790, Portsmouth History Centre, Vignoles Papers 1072A/App. 1 (printed in Keith H. Vignoles, *The Infant Ensign: the story of Charles and Camilla Vignoles and their son Charles Blacker* (Emsworth, 1967) as appendix 1; see also appendix 2); marriage licence of Camilla Hutton and Cha[rle]s Henry Vignoles, dated 16 December 1790 accessed via www.findmypast. co.uk; Olinthus J. Vignoles, *Life of Charles Blacker Vignoles: a reminiscence of early railway history* (London, 1889), 4; Keith Hutton Vignoles, *Charles Blacker Vignoles, romantic engineer* (Cambridge, 1982), 1.

147 *the birth of his grandson*: Letter of Charles Henry Vignoles to Charles Hutton, 3 Jun 1793, Portsmouth History Centre, Vignoles Papers 1072A/App. 4 (printed in Vignoles, *Infant Ensign* as appendix 4).

147 *ordered to the West Indies*: K.H. Vignoles, *Charles Blacker Vignoles*, 1; Kennedy, *Narratives of War*, 104.

9 Utility and Fame

148 *The twenty-seventh of September, 1794*: May, *Charlton*, 69.

148 *We commit her body*: *The Book of Common Prayer* (London, edition of 1794), 376.

148 *Two brief biographies*: *Public Characters*; 'Biographical Anecdotes'.

148 *'distinguished', 'learned', even 'veteran'*: Review of Hutton, *Tracts* (1812) in *The Critical Review* (January and February 1814), 1–13 and 109–123 at 1.

149 *Hutton's opinion about bridge-building projects*: letter of Charles Hutton to R. Benson Esquire, 18 Jun 1792, Plymouth and West Devon Record Office, 105/164; letter of Charles Hutton to Robert Harrison, between Sep 1794 and Aug 1795, printed in Melmore, 'Some Letters', 79–80; letter of Charles Hutton to W. Rennie, 10 Feb 1802, National Archive, PRO 30/9/131, fos. 20–22; Charles Hutton, manuscript reply to queries about London Bridge, 18 Jun 1819, Portsmouth History Centre, Vignoles papers, letter 186/1; Hutton, *Tracts* (1812), vol. 1, pp. 127–44.

149 *aimed not at fame*: review of Hutton, *Tracts* (1812) in *The Quarterly Review* (1813), 400–18; Gregory, 'Memoir', 226.

150 *an agreement with Joseph Johnson*: University of Cincinnati MS Q121.H93 1786: Memorandum of agreement between Charles Hutton and Jos. Johnson, 20 May 1786.

150 *fondness for French culture*: Hutton, *Tables* (1785), 41; Hutton, *Dictionary*, vol. 1, v.

151 *the article on algebra*: *Public Characters*, vol. 2, p. 121; Trinity College Cambridge MS R.1.59, 64r–66r, 72r–74r (Hutton's notes on mathematical books of the sixteenth century, and on the history of algebra); London, Senate House Library MS 235 (Hutton's translation from Tartaglia).

152 *biographies of ancient and modern mathematicians*: 'Memoir of the Late Dr. Hutton', *The Gentleman's Magazine* (March 1823), 228–32 at 228.

152 *A letter from Nevil Maskelyne to Hutton*: Nevil Maskelyne to Charles Hutton, 20 June 179[ ], Cambridge University Library, RGO 35/92 (photocopy).

152 *Hutton mentioned for instance . . .*: Hutton, *Dictionary*, vol. 1, pp. 20, 67, 557; letter of John Playfair to Charles Hutton, 21 Apr 1788, Wellcome Collection MS 7430 no. 38; letter of Charles Hutton to Francis Baily, 13 Jul 1808, London, Senate House Library, [DeM] L.4 [Waring] (copy of Edward Waring, *On the Principles of Translating Algebraic Quantities* (Cambridge, 1792); letter pasted in at end).

153 *'Learned Societies throughout Europe'*: Hutton, *Dictionary*, vol. 1, p. vi.

153 *'just got to the end'*: Letter of Charles Hutton to Nevil Maskelyne, 19 Dec 1793, Cambridge University Library, RGO 4/187/18; letter of Charles Hutton to Robert Harrison, between Sep 1794 and Aug 1795, printed in Melmore, 'Some Letters', 79–80; Hutton, *Dictionary*, vol. 1, p. vi.

153 *a very small typeface*: letter of Charles Hutton to Robert Harrison, between Sep 1794 and Aug 1795, printed in Melmore, 'Some Letters', 79–80.

154 *initially issued in separate numbers*: advert for instance in *St. James's Chronicle*, 24 January 1795.

154 *a single-leaf puff*: *Proposals for Publishing a Mathematical and Philosophical Dictionary* [London, n.d.]; Cambridge University Library, White b.8, a copy of the *Proposals* pasted to a letter of Charles Hutton to David Stephenson, 7 February 1795.

155 *a good deal of anticipation*: letter of John Playfair to Charles Hutton, 25 Aug 1792, Wellcome Collection MS 5270 no. 44; letter of David Kinnebrook to his father, 23 Feb 1795, Cambridge University Library, RGO 35/106; advert for instance in *St. James's Chronicle*, 4 February 1796.

155 *'smatterers and would-be scholars'*: Review of Hutton, *Dictionary* in *English Review* 28 (July 1796), 14–19 at 18; David Rivers, *Literary Memoirs of Living Authors of Great Britain* (London, 1798), 300; Review of Margaret

Bryan, *A Compendious System of Astronomy* (London, 1797) in *The Monthly Visitor* 8 (1799), 93–9 at 93.

155 *'of an equal and uniform nature'*: Hutton, *Dictionary*, vol. 1, p. vii.

156 *Charlotte . . . ruptured a vessel*: Obituaries of Charlotte Hutton in *The Gentleman's Magazine* (October 1794), 960–61; in the *Hibernian Magazine* (1794), 477–8; 'Hutton, Charlotte' in Johnson and Exley, *Imperial Encyclopaedia*, vol. 3, 60–61.

156 *'gloomy remainder'*: Charles Hutton to Robert Harrison, between Sep 1794 and Aug 1795, printed in Melmore, 'Some Letters', 79–80.

157 *I dreamt that I was dead*: Obituary of Charlotte Hutton in *The Gentleman's Magazine*, 961.

158 *David Kinnebrook*: Letter of David Kinnebrook to his father, 23 Feb 1795, Cambridge University Library, RGO 35/106.

158 *Henry had heard the truer word*: K.H. Vignoles, *Charles Blacker Vignoles*, 2–3; Vignoles, *Infant Ensign*, 9–11; obituary of Henry Hutton in *The Gentleman's Magazine* (December 1827), 561–2; Charles Hutton to Robert Harrison, 13 Aug 1795, printed in Melmore, 'Some Letters', 80–81.

160 *'I had them given a decent burial'*: M. Courtois to Charles Hutton, 15 December 1794 (25 Frimaire, an 3), Portsmouth History Centre, Vignoles Papers 1072A/App. 10 (printed in Vignoles, *Infant Ensign* as appendix 10).

160 *a section here and a section there*: Hutton, *Course of Mathematics*, vol. 1, p. iii.

161 *a hundred guineas*: Jones, *Records of the RMA*, 46; also 51. Oxford, Bodleian Library, MS. Eng. misc. b. 190, fos. 1–2 – the articles of agreement between Charles Hutton and George and John Robinson for the publication of the *Course* – values a half share in the *Course* at £125.

161 *Every cadet had to buy*: Jones, *Records of the RMA*, 46.

162 *'Suppose 471 men are formed'*: Hutton, *Course*, vol. 1, p. 20.

162 *mixed reviews*: Reviews of Hutton, *Course* in *The Critical Review* 25 (February 1799), 159–62 at 161; in *The New London Review* 4 (April 1799), 342–5 at 342.

162 *he shall I name*: Vibart, *Addiscombe*, 118, from a poem titled 'Addiscombe: a tale of our times' by J.H. Burke, printed in 1834.

163 *he may have corresponded with Montucla*: Jean E. Montucla and Joseph Jérôme L. de Lalande, *Histoire des Mathématiques* (Paris, 1799–1802), vol. 3, p. 108 and Niccolò Guicciardini, personal communication; Hutton's interest in Newtoniana makes him probably the best candidate to be the correspondent mentioned. Hutton, *Dictionary*, vol. 2, p. 157; letter of Mark Noble to Rev. Mark Noble, 5 July 1809, Bodleian MS Eng. misc. d. 160, fos. 181–2.

164 *the impossibility of determining*: Charles Hutton, *Recreations in Mathematics and Natural Philosophy* (4 vols: London, 1803), vol. 4, p. 47.

164 *a project cooked up in collusion*: Review of Hutton, *Dictionary* in *English Review* 28 (July 1796), 14–19 at 18.

165 Select Amusements in Philosophy and Mathematics: M.L. Despiau, *Select Amusements in Philosophy and Mathematics* (London, Edinburgh and Glasgow, 1801); obituaries of Hutton in *The European Magazine* (June 1823), 483–7 at 487; in the *Literary Gazette* (1 February 1823), 75–6 at 75.

165 'warm commendation': Review of Hutton, *Recreations* in *The British Critic* 21 (May 1803), 542–51 at 551.

165 'well adapted to lie': Review of Hutton, *Recreations* in *Monthly Review* 44 (May 1804), 35–40 at 40.

165 'The whole work indeed reflects much credit': Review of Hutton, *Recreations* (1814 edition) in *The British Critic* 63 (April 1815), 409–18 at 418.

165 'insipid and puerile': Review of Hutton, *Recreations* in *The Imperial Review* 4 (January and February 1805), 3–12 and 155–166, at 4.

166 *some were scarcely to be had*: Hutton et al., *Philosophical Transactions Abridged*, vol. 1, pp. v–vi.

166 *An attempt was announced in 1802*: *Prospectus of an Abridgment of the Philosophical Transactions* (London, Bunney and Gold, 1802) (two copies in the John Johnson collection of printed ephemera).

166 *a short prospectus*: *Prospectus of a New Abridgement of the Philosophical Transactions* (London, 1803); *Prospectus of the New Abridgement of the Philosophical Transactions of the Royal Society of London* (two forms: London, 1806) (copies in the John Johnson collection of printed ephemera); advert (separately paginated) in *The Gentleman's Magazine* (May 1809).

166 *the* Abridgement *was under Hutton's principal care*: Gregory, 'Memoir', 216; obituary of Hutton in *The Edinburgh Annual Register* 16 (December 1823), 328–31 at 330.

167 *that colossal figure*: for comparison, the assignments of copyright in *Original Letters, Manuscripts, and State Papers, Collected by William Upcott* ([London] 1836), 50, show prices ranging from £284 for a half share in Hutton's *Mensuration, Arithmetic* and *Diarian Repository* and £105 for the *Logarithms*, down to £31 for the *Key to Hutton's Arithmetic*.

167 *Annotations on papers*: Hutton et al., *Philosophical Transactions Abridged*, vol. 14, p. 420 (Hutton on the density of the earth); vol. 14, p. 551 (Ingenhousz on inflammable air and gunpowder); vol. 15, pp. 102, 103, 107 (Thompson on gunpowder); vol. 18, 142–177 (Cavendish on the density of the earth).

167 *a subject classification*: Review of Hutton et al., *Philosophical Transactions Abridged* vols 1 and 2 in *Literary Journal* (1804), 259–62 at 261–2; Hutton et al., *Philosophical Transactions Abridged*, vol. 3, pp. v–xiii; Maurice Crosland and Crosbie Smith, 'The Transmission of Physics from France to Britain: 1800–1840', *Historical Studies in the Physical Sciences* 9 (1978), 1–61 at 4.

167 *Hutton wrote to Banks*: Royal Society, CMO/8, 218 (council meeting of 31 March 1803).

168 *to dedicate the work*: Hutton et al., *Philosophical Transactions Abridged*, vol. 1, [A]2r.

168 *a planned supplementary volume*: Prospectus of the New Abridgement (1806: longer form), 3; Hutton *et al.*, *Philosophical Transactions Abridged*, vol. 1, p. 319.

168 *criticisms from* The Monthly: Review of Hutton et al., *Philosophical Transactions Abridged*, vols 1–3 in *Monthly Review* (1805), 284–6 (quote 286).

168 *The* British Critic: Review of Hutton et al., *Philosophical Transactions Abridged*, vol. 1 in *The British Critic* (1803), 538–44 at 544; Review of Hutton et al., *Philosophical Transactions Abridged*, vols 1–2 in the *Literary Journal* (1804), 259–62 at 259; notice of Hutton et al., *Philosophical Transactions Abridged*, vols 1–4 in Thomas Leybourn, *Mathematical Repository*, new series, vol. 1 (1799), 135; quotation from the *Journal de la littérature étrangère* 1 (An 13), 37 in *Prospectus of the New Abridgement* (1806: longer form), 4.

10 Securing a Legacy

171 *Mount Schiehallion*: John Playfair, 'Account of a Lithological Survey of Schehallien, made in order to determine the specific gravity of the rocks which compose that mountain', *Philosophical Transactions* 101 (1811), 347–77 at 348–55.

172 *The Academy at last moved*: Shepperd, *Sandhurst*, 33.

172 *The Cube House was purchased*: Jones, *Records of the RMA*, 55.

172 *an unexpected embarrassment*: Charles Hutton, 'Note on the Divining Rod', *Philosophical Magazine* 55 (June 1820), 465–8 at 467; Saint and Guillery, 'Woolwich Common', 25.

172 *he was granted permission*: Jones, *Records of the RMA*, 57.

173 *His new town house*: Letter of Charles Hutton to Messrs Kent, 16 Nov ?1807, Oxford, Bodleian Library, MS Montagu d. 7, fol. 562 (Hutton's first extant letter addressed from 34 Bedford Row).

173 *a small fortune*: Letter of Charles Blacker Vignoles to Mary Vignoles, 9 Jul 1823, Portsmouth History Centre, Vignoles Papers, letter 286.

173 *two pianos*: Letters of Charles Blacker Vignoles to Mary Vignoles, 28 Jun and 4 Aug 1823, Portsmouth History Centre, Vignoles Papers, letters 282 and 295.

173 *raise subscriptions and build a new theatre*: Notice of subscription for a new theatre in *Statesman* (1 November 1809).

the official table: Hutton, *Dictionary*, vol. 2, p. 728.

174 *often asked to recommend mathematicians*: Bruce, *Memoir*, 37; *Tribute of Respect to Charles Hutton, LL.D. F.R.S. &c. &c.* [London, 1822], 2.

174 *Lewis Evans*: Correspondence of Lewis Evans and Charles Hutton, Oxford, Museum of the History of Science, MSS Evans 31; Jones, *Records of the RMA*, 49.

174 *David Kinnebrook*: letters of David Kinnebrook to his father, 23 Feb and 29 Oct 1795, Cambridge University Library RGO 35/106 and 35/115; Croarken, 'Longitude for All', 119; Croarken, 'Astronomical Labourers', 289.

174 *Charles Wildbore*: Letter of Charles Hutton to Robert Harrison, 31 May 1780, printed in Melmore, 'Some Letters', 77–8.

174 *Edward Riddle*: Bruce, *Memoir*, 37; Sykes, *Local Records* (1867), 288–9.

174 *John Bonnycastle*: Gregory, 'Memoir', 226; Jones, *Records of the RMA*, iii, iv, 36, 42, 44, 49, 51, 56, 82.

174 *Margaret Bryan*: Margaret Bryan, *A Compendious System of Astronomy* (London, 1797), x–xi.

174 *Major Edward Williams and Lieutenant William Mudge*: Gregory, 'Memoir', 226; W.A. Seymour and J.H. Andrews, *A History of the Ordnance Survey* (Folkestone, 1980) 12, 15; C.F. Close, *The Early Years of the Ordnance Survey* (London, 1926), 30.

174 *Sir John Leslie*: Letters of Sir John Leslie to Charles Hutton, 14 Oct 1795, and of Charles Hutton to Joseph McCormack, ?19 Oct 1795, London, Senate House Library, MS 913B/3/1 (xv: letter and draft reply); also of Charles Hutton to Edinburgh Town Council, 19 Feb 1805, Edinburgh City Archives, SL1/1/142: Town Council Minutes vol. 142, pp. 139–141 (copy); see J.B. Morrell, 'The Leslie Affair: careers, kirk and politics in Edinburgh in 1805', *The Scottish Historical Review* 54 (1975), 63–82 at 77.

175 *Charles Babbage*: Note on the mathematical chair at Edinburgh, *Caledonian Mercury* (2 September 1819); Charles Hutton, testimonial for Charles Babbage, 21 March 1816, British Library, Add. MS 37182, fos. 60–61.

175 *'serving and encouraging'*: Gregory, 'Memoir', 226–7, quoting from Hutton's (now lost) manuscript journal for 1821.

175 *a quite extraordinary fracas*: Letters of Lewis Evans and of the Trustees of the Grammar School at Hungerford to Charles Hutton, 29 Sep 1788, Oxford, Museum of History of Science, MS Evans 31 (drafts).

175 *Olinthus Gregory*: Alexander Gordon, rev. Ben Marsden, 'Gregory, Olinthus Gilbert (1774–1841), mathematician' in *ODNB*.

176 *When Charles Wildbore died*: Albree and Brown, 'A Valuable Monument', 26. Despite my assertions in *Poor Robin's Prophecies: a curious Almanac, and the everyday mathematics of Georgian Britain* (Oxford, 2012), close examination of the evidence makes it seem that the editorship of *Old Poor Robin* remained linked to that of the *Gentleman's Diary*, for which Hutton twice recommended candidates but which he never held himself.

176 *second assistant mathematics master*: Jones, *Records of the RMA*, iv, 53, 55, 82.

176 *work with ballistic pendulums*: Essay review including Hutton, *Tracts* (1812), in *The British Review* 20 (1822), 283–300, at 287, 291; Douglas, *Naval Gunnery*, 28, 39, 92.

177 *Gregory took on both*: Gordon, 'Gregory'.

178 *'liberal encouragement of the Public'*: Hutton, *Tracts* (1812), vol. 1, p. x.

178 *'held in high estimation'*: Obituary of Hutton in *The Edinburgh Annual Register* 16 (December 1823), 328–31, at 328.

178 *'an inaccurate transcript'*: Letter of Lewis Evans to Charles Hutton, 26 May 1789, Oxford, Museum of the History of Science, MS Evans 31: draft.

178 *His* Course *was adopted*: V. Frederick Rickey and Amy Shell-Gellasch, 'Mathematics Education at West Point: the first hundred years – Charles Hutton': http://www.maa.org/publications/periodicals/convergence/mathematics-education-at-west-point-the-first-hundred-years-charles-hutton; Frank J. Swetz, 'The Mystery of Robert Adrain', *Mathematics Magazine* 81 (2008), 332–44 at 340; Hutton, *Course* (New York, 1812), xi.

178 *West Point also had a copy*: *Catalogue of Books in the Library of the Military Academy* (United States Military Academy, 1822), 17.

179 *an unknown translator*: Ann Arbor, MI, University of Michigan, Mich. Ms. 276 (UF 144.H49), 'Expériences de pyrotechnie d'Hutton', a French translation of Hutton's 'New Experiments in Artillery . . .' from *Tracts* (1786), dated 1791 or 1792; attribution in later endorsement on first leaf ('traduction privée inédite faite à la demand du chimiste Guyton de Morveau').

179 *Joseph-Louis Lagrange was working*: Louis-Bernard Guyton de Morveau, 'Notes', appended to J.B.J. Delambre, 'Notice sur la vie et les ouvrages de M. Malus, et de M. le Comte Lagrange', *Mémoires de la classe des sciences mathématiques et physiques de l'institut impérial de France. Année 1812. Première partie* (Paris, 1814), lxxviii–lxxx, at lxxix; obituary of Hutton in *Monthly Magazine* 55 (March 1823), 137–42 at 139.

179 *a French colonel named Villantroys*: Charles Hutton, *Nouvelles expériences d'artillerie*, tr. Pierre-Laurent de Villantroys (Paris, 1802).

179 *Charles Dupin included an illustration*: Essay review including Hutton, *Tracts* (1812), in *The British Review* 20 (1822), 283–300 at 287–8, 298.

180 *A French reviewer*: J. Madelaine, review of Hutton, *Nouvelles expériences d'artillerie* in *Journal des sciences militaires* 5 (1826), 350–79 at 354.

180 *Alexander Ingram in 1796*: Charles Hutton, *A Complete Treatise on Practical Arithmetic . . . A new edition, corrected and enlarged, by Alexander Ingram* (London, 1796); Daniel Dowling, *Key to the Course of Mathematics, composed . . . by Charles Hutton* (London, 1818); J.M. Edney, *A Key to Hutton's Compendious Measurer* (London, 1824).

181 *the* Course *in particular functioning*: Jenny Bulstrode, 'The Promiscuous Exercises of the Woolwich Bois Boys', paper given at All Souls College, Oxford, December 2015.

181 *'Any hints of improvements'*: Hutton, *Key* (1786), [A]2v.

181 Diary *contributors, in particular*: Letter of Lewis Evans to Charles Hutton, 29 Sep 1788, Oxford, Museum of the History of Science, MS Evans 31 (draft); letter of Charles Hutton to Robert Harrison, 13 Aug 1795, printed in Melmore, 'Some Letters', 80–81; review of Hutton, *Mensuration* (1788) in *The Critical Review* 67 (March 1789), 207–10 at 207.

181 *a long and eccentric correspondence*: Letters of Lady M. to Charles Hutton, 10 Feb, 30 Apr and 23 May 1805, and 30 Oct 1813, Hutton, *Recreations* (1814), vol. 4, pp. 223–9, 231; 'The Divining Rod', *Newcastle Magazine* 2 (January 1823), 48–50; letter to the editor in the *Monthly Magazine* (July 1820), 515–17.

181 *a third volume of the* Course: Hutton, *Course* (1811 edition), vol. 3, pp. iii–iv.

182 *'may best be used in tuition'*: Hutton, *Course* (1811 edition), vol. 3, p. viii.

182 *Cavendish's was a beautiful experiment*: Henry Cavendish, 'Experiments to Determine the Density of the Earth', *Philosophical Transactions* 88 (1798), 469–526.

183 *Hutton never accepted Cavendish's result*: Letter of Charles Hutton to Henry Cavendish, 17 Nov 1798, Devonshire Collections, Chatsworth: Cavendish MSS New Correspondence, printed in Christa Jungnickel and Russell McCormmach, *Cavendish: the experimental life* (2nd edn, s.l., 1999), 710–11; Letter of Charles Hutton to the Marquis de Laplace, 9 Apr 1819, printed in the *Philosophical Magazine* ser. 1, 55 (February 1820), 81–5 and (in French) in the *Journal de Physique, de chimie, et de l'Histoire Naturelle* 90 (April 1820), 307–12.

183 *In 1778 he had assumed*: Hutton, 'Account of the Calculations', 782; Charles Hutton, 'Calculations to Determine at What Point in the Side of a Hill its Attraction will be the Greatest', *Philosophical Transactions* 70 (1779), 1–14; Hutton et al., *Philosophical Transactions Abridged*, vol. 14, p. 420; Hutton, *Dictionary* (1815), vol. 1, p. 403.

184 *in 1801 Playfair carried out*: Playfair, 'Account of a Lithological Survey'; Hutton, *Tracts* (1812), vol. 2, p. 63.

184 *'greater than can easily be imagined'*: Reviews of Hutton, *Tracts* (1812) in *The Edinburgh Review* 22 (October 1813), 88–107 at 96; and in *The Quarterly Review* 9 (July 1813), 400–18 at 407.

185 *trigonometric tables in radians*: Charles Hutton, 'Project for a New Division of the Quadrant', *Philosophical Transactions* 74 (1784), 21–34; Hutton, *Tables* (1785), x (text retained in the edition of 1794 though not in those of 1801 and subsequently), 2–3; letter of Lewis Evans to Charles Hutton, 26 May 1789, Oxford, Museum of the History of Science, MSS Evans 31; Hutton, *Tracts* (1812), vol. 2, p. 133; Gregory, 'Memoir', 213.

185 *the Royal Society's librarian*: Trinity College, Cambridge, MS R.1.59, fos. 54–63; see also Hutton, *Tracts* (1812), 115–26.

185 *selections and abridgements*: Trinity College, Cambridge, MS R.1.59, fos. 179–217 (quote fol. 179r), a mixture of material from Hutton and Gregory.

185 *a book-length translation*: London: Senate House Library, MS 235 (translation from Niccolò Tartaglia, *Quesiti, et inuentioni diverse* (Venice, 1546), book 9).

186 *a few new results*: Hutton, *Tracts* (1812), vol. 1, pp. 260–65; vol. 2, pp. 122–40.

186 *new material on the history of algebra*: Hutton, *Tracts* (1812), vol. 2, pp. 151–64, 195–201; Trinity College, Cambridge, MS R.1.59, fos. 124–153; see also Ivahn Smadja, 'On Two Conjectures that Shaped the Historiography of Indeterminate Analysis: Strachey and Chasles on Sanskrit sources', *Historia Mathematica* 43 (2016), 241–87.

186 *a history of iron bridges*: Hutton, *Tracts* (1812), vol. 1, pp. 144–66.

186 *the final account of the work on ballistics*: Hutton, *Tracts* (1812), vol. 2, pp. 306–84 and vol. 3.

187 *a final legacy*: Hutton, *Tracts* (1812), vol. 1, pp. iii, x.

187 *to modify some of his judgements*: Hutton, *Tracts* (1812), vol. 2, pp. 13, 38; within the paper on the density of the earth credits to Maskelyne and Cavendish were removed. See above on Hutton's alteration of the concluding result in this paper.

187 *compiled and ready by July 1812*: Hutton, *Tracts* (1812), vol. 1, p. x; vol. 2, pp. 166–93 (the EIC librarian at 167).

188 *some long and prominent notices*: Reviews of Hutton, *Tracts* (1812) in *The Quarterly Review* 9 (July 1813), 400–18; *The Edinburgh Review* 22 (October 1813), 88–107 (quote p. 88; Playfair identified as author in *The Wellesley Index to Victorian Periodicals*, wellesley.chadwyck.com); *The Critical Review* 5 (January, February and April 1814), 1–13, 109–23 and 374–81; *The New Annual Register* (January 1814), 317–29; *The British Review* 20 (December 1822), 283–300.

11 Controversies Old and New

189 *Till within the last year*: letter of Charles Blacker Vignoles to Mary Griffiths, 8 May 1814, Portsmouth History Centre, Vignoles Papers, Letter 65 (appendix).

192 *'disgraceful management'*: Saint, *Four Letters*, iv, 7.

192 *A profusion of new orders*: Saint, *Four Letters*, 17, 63–5; Jones, *Records of the RMA*, 57–61.

192 *nearly two hundred cadets*: Saint, *Four Letters*, 59–60.

192 *indulging in games*: Saint, *Four Letters*, 78, 28.

192 *Hutton's Course remained the foundation*: Jones, *Records of the RMA*, 60, 69, 82, 96, 102 (use of Hutton's *Course* discontinued at the RMA in 1840).

193 *A distinctly mixed review*: review of Hutton, *Dictionary* in *Monthly Review* 25 (February and April 1798), 184–201 and 364–83, at 190.

193 *a frankly rude one-page notice*: Review of Hutton, *Bridges* (1801) in *Monthly Review* 37 (March 1802), 323–4 at 324.

193 *Robert Woodhouse*: B.C. Nangle, *The Monthly Review, Second Series, 1790–1815: indexes of contributors and articles* (Oxford, 1955), vol. 2, p. 74; letter of Charles Hutton to W. Windham, 15 Sep 1802, British Library, Add. MS 37915, fol. 218–19.

193 *Hutton wrote in some wrath*: Letter of Charles Hutton to Ralph Griffiths, 1 May 1802, Oxford, Bodleian Library, MS Add. C. 89, fol. 167 (this is the third letter in the sequence, Hutton's first and Griffiths's reply being lost).

193 *a short essay on the 'abuse of reviews'*: Charles Hutton, 'To the Editor of the Monthly Magazine', *Monthly Magazine* (August 1802), 27–31.

194 *a shorter piece in the* Monthly Review: Charles Hutton, 'To the Editor of the Monthly Review', *Monthly Review* 38 (June 1802), 222–4.

194 *lukewarm reviews*: Reviews of Hutton, *Recreations* in *Monthly Review* 44 (May 1804), 35–40; and of Hutton et al., *Philosophical Transactions Abridged* in *Monthly Review* 47 (July 1805), 284–6 (quote at 286).

194 *'a general propensity'*: Letter of Charles Hutton to Ralph Griffiths, 1 May 1802, Oxford, Bodleian Library, MS Add. C. 89, fol. 167.

194 *the distinctive British methods*: Niccolò Guicciardini, 'Dot-Age: Newton's mathematical legacy in the eighteenth century', *Early Science and Medicine* 9 (2004), 218–56; R.E. Schofield, 'An evolutionary taxonomy of eighteenth-century Newtonianisms', *Studies in Eighteenth-Century Culture* 7 (1978), 175–92; Guicciardini, *Development*.

194 *'A certain degree'*: [John Playfair], review of P.S. Laplace, *Traité de mécanique céleste*, vols 1–4 (Paris, 1799–1805) in *Edinburgh Review* 11 (January 1808), 248–84 at 282.

194 *'By means of such problems'*: Hutton, *Miscellanea Mathematica*, 1.

195 *France, and the European continent*: Crosland and Smith, 'The Transmission of Physics', 6–7; Guicciardini, *Development*, 141; Guicciardini, 'Dot-Age', 253–5.

196 *he had continental mathematics books*: A Catalogue of the . . . Library of Charles Hutton, passim.

196 *some published translations*: Guicciardini, *Development*, 108–13, 119; Florian Cajori, 'Discussion of Fluxions: From Berkeley to Woodhouse', *The American Mathematical Monthly* 24 (1917), 145–54; Leybourn, *Mathematical Repository*, new series (1799–1814) contained translations of works by Lagrange, Legendre and Euler by various authors. Alex D.D. Craik, 'Mathematical Analysis and Physical Astronomy in Great Britain and Ireland, 1790–1831: some new light on the French connection', *Revue d'histoire des mathématiques* 22 (2016), 223–94.

196 *perhaps constrained*: Guicciardini, *Development*, 108.

196 *Laplace published*: Crosland and Smith, 'The Transmission of Physics', 17; John Toplis, *A Treatise Upon Analytical Mechanics; being the first book of the Mécanique céleste* (Nottingham, 1814); Guicciardini, *Development*, 117; Jonathan R. Topham, 'Science, Print, and Crossing Borders: importing French science books into Britain, 1789–1815', in David N. Livingstone and Charles Withers (eds), *Geographies of Nineteenth-Century Science* (Chicago, 2011), 311–44 at 326–8; also Playfair's review cited above.

196 *During the last half century*: Review of Toplis, *Analytical Mechanics* in *Monthly Review* 78 (October 1815), 211–13 at 211, 212.

197 *A man may be perfectly acquainted*: [John Playfair], review of P.S. Laplace, *Traité de mécanique céleste*, vols 1–4 (Paris, 1799–1805) in *Edinburgh Review* 11 (January 1808), 248–84 at 323–4; Guicciardini, *Development*, 102; Amy Ackerberg-Hastings, 'John Playfair on British Decline in Mathematics', *BSHM Bulletin: Journal of the British Society for the History of Mathematics* 23 (2008), 81–95.

197 *the short-lived Analytical Society*: Warwick, *Masters of Theory*, 67–8; Philip C. Enros, 'The Analytical Society (1812–1813): precursor of the renewal of Cambridge mathematics', *Historia Mathematica* 10 (1983), 24–47; Guicciardini, *Development*, 135–6.

197 *much of their rhetoric*: Elizabeth Garber, 'On the Margins: experimental philosophy and mathematics in Britain, 1790–1830' in *The Language of Physics: the calculus and the development of theoretical physics in Europe, 1750–1914* (Boston, 1999), 169–206, at 191.

197 *British mathematics was now being transformed*: Guicciardini, 'Dot-Age', 256; Crosland and Smith, 'The Transmission of Physics', 18, 59–60 and *passim*; Guicciardini, *Development*, 128–30, 138–9.

198 *Gregory himself was not wholly unsympathetic*: Albree and Brown, 'A Valuable Monument', 26; but compare the view expressed in Hutton, *Course* (1836–7), vol. 2, p. 203.

198 *There was a period*: O.J. Vignoles, *Charles Blacker Vignoles*, 9.

198 *Vignoles acquired a grounding*: K.H. Vignoles, *Charles Blacker Vignoles*, 5; Gilbert Austin to Charles Hutton, 13 Aug 1806, Portsmouth History Centre, Vignoles Correspondence, 1072A/O1.

199 *'too fond of a dash'*: K.H. Vignoles, *Charles Blacker Vignoles*, 11.

199 *working through logarithmic and other tables*: O.J. Vignoles, *Charles Blacker Vignoles*, 9.

199 *attempting to document the sums*: Portsmouth History Centre, Vignoles Papers 1072A/App. 15 (administration of Charles Henry Vignoles, with an account of expenses), App. 13 (draft letter of Charles Hutton to W. Windham, 20 Feb 1797).

199 *rumours about grants of land*: Letter of John Vignoles to Charles Hutton, 21 Oct 1814, Portsmouth History Centre, Vignoles Papers, Letter 75/1.

200 *He was placed under articles*: K.H. Vignoles, *Charles Blacker Vignoles*, 6; letter of Margaret Hutton to Charles Blacker Vignoles, 3 Aug 1816, Portsmouth History Centre, Vignoles Papers, Letter 122.

200 *quite desperate for escape*: Letter of Charles Blacker Vignoles to Mary Griffiths, 8 May 1814, Portsmouth History Centre, Vignoles Papers, Letter 65 ('under the eye of such a Man'); also repeated references to 'independence' in later letters such as no. 138 (15 Jun 1817), no. 172 (17 May 1818), no. 261 (25 Mar 1823).

200 *an incident with a girl*: O.J. Vignoles, *Charles Blacker Vignoles*, 31; K.H. Vignoles, *Charles Blacker Vignoles*, 6.

200 *Vignoles left Bedford Row*: Letter of Margaret Hutton to Charles Blacker Vignoles, 11 Aug 1814, Portsmouth History Centre, Vignoles Papers, Letter 72.

200 *private tuition from Leybourn*: O.J. Vignoles, *Charles Blacker Vignoles*, 14; K.H. Vignoles, *Charles Blacker Vignoles*, 7.

200 *Hutton wasn't altogether happy*: Mary Vignoles to Charles Blacker Vignoles, 15 Feb 1823, Portsmouth History Centre, Vignoles Papers, Letter 258.

200 *military adventures*: K.H. Vignoles, *Charles Blacker Vignoles*, 9–12.

201 *Margaret was desperately worried*: Letters of Margaret Hutton to Charles Blacker Vignoles, 11 Aug 1814, and of Charles Hutton Jr and Isabella Hutton to Charles Blacker Vignoles, 22 Aug and 28 Sep 1814, Portsmouth History Centre, Vignoles Papers, Letters 72 and 75.

201 *a manuscript tragedy*: O.J. Vignoles, *Charles Blacker Vignoles*, 32, 43–4, 57 (quote); K.H. Vignoles, *Charles Blacker Vignoles*, 6, 12.

201 *he was put on half pay*: O.J. Vignoles, *Charles Blacker Vignoles*, 50; Anthony Hall-Patch, 'Charles Blacker Vignoles, F.R.S.', *Notes and Records of the Royal Society of London* 47 (1993), 233–42 at 234.

202 *Henry's son Charles*: May, *Charlton*, 69; parish register for St Mary Magdalene, Canterbury, accessed via ancestry.co.uk (baptism on 8 January 1800).

202 *In August he wrote home proudly*: Letters of Margaret Hutton to Charles Blacker Vignoles, 14 Mar 1814, and of Charles Hutton Jr and Isabella Hutton to Charles Blacker Vignoles, 22 Aug and 28 Sep 1814, Portsmouth History Centre, Vignoles Papers, Letters 38 and 75.

202 *he sickened of the life*: Letters of Margaret Hutton to Charles Blacker Vignoles, 11 Aug and 21 Nov 1814, Portsmouth History Centre, Vignoles Papers, Letters 72 and 76/1 (quote); O.J. Vignoles, *Charles Blacker Vignoles*, 37.

202 *In October he was moved*: Letter of Isabella Hutton to Charles Blacker Vignoles, 24 Mar 1815, Portsmouth History Centre, Vignoles Papers, Letter 80.

202 *his creative scientific work was over*: A Catalogue of the Library . . . of Charles Hutton, title page; letter of Charles Hutton to John Bruce, 1815, extract printed in Bruce, *Memoir*, 32–3 (quote).

203 *the best in Britain*: letter of Charles Hutton to John Bruce, 1815, extract

printed in Bruce, *Memoir*, 32–3; *A Catalogue of the Library . . . of Charles Hutton*, *passim*; *Public Characters*, vol. 2, pp. 100–01, 103, 112.

204 *implacable*: letter of Charles Hutton to John Bruce, 1815, extract printed in Bruce, *Memoir*, 32–3; [Gregory], 'A Review', 248.

204 *Isabella first worked through the library*: New Haven, CT, Yale University, Lewis Walpole Library, LWL Mss Vol. 54 (2 vols), Catalogue of Charles Hutton's library, 1816.

204 *greeted with horror*: Bruce, *Memoir*, 32–3.

204 *twice the size of Playfair's*: Jungnickel and McCormmach, *Cavendish*, 323.

204 *Friends of Hutton acquired*: Eric G. Forbes, 'Collections II: the Crawford Collection of books and manuscripts on the history of astronomy, mathematics, etc., at the Royal Observatory, Edinburgh', *The British Journal for the History of Science* 6 (1973), 459–61 at 459–60; Bruce, *Memoir*, 33; Wallis, 'Mathematical Tradition', 11, 41.

205 *A London bookseller named Weale*: *A Catalogue of Books, on the Sciences: astronomy, mathematics, natural philosophy, &c. with some added that are curious and miscellaneous, chiefly from the libraries of Rev. Nevil Maskelyne . . . Bishop Horsley . . . Dr. Charles Hutton . . . William Phillips . . . and Richard Heber . . . On sale, by John Weale . . . 59, High Holborn, London* [London, 1835] (originally issued with *London and Edinburgh Philosophical Magazine*, series 3, no. 6 (April 1835)).

205 *mathematical instruments of Benjamin Franklin*: *A Catalogue of the Library . . . of Charles Hutton*, 80.

206 *lamenting the sale*: Letters of Charles Hutton to John Bruce, 8 May 1817 and 22 Mar 1822, printed in Bruce, *Memoir*, 33–4, 39–42.

206 *Margaret Hutton, too, sickened*: Letters of Margaret Hutton to Charles Blacker Vignoles, 11 Aug 1814 and 13 Dec 1816, Portsmouth History Centre, Vignoles Papers, Letters 72 and 126; May, *Charlton*, 69.

207 *'Let it not be said'*: Letter of Margaret Hutton to Charles Blacker Vignoles, 11 Aug 1814 and 13 Dec 1816, Portsmouth History Centre, Vignoles Papers, Letter 72.

207 *'almost alone'*: Letter of Charles Hutton to John Bruce, 8 May 1817, printed in Bruce, *Memoir*, 33–4.

207 *a cook and a maid*: Letters of Isabella Hutton to Charles Blacker Vignoles, 29 Jul 1830 and 27 Dec 1830, Portsmouth History Centre, Vignoles Papers, Letters 498 and 514.

207 *An attractive ward*: K.H. Vignoles, *Charles Blacker Vignoles*, 8.

207 *a voluminous clandestine correspondence*: K.H. Vignoles, *Charles Blacker Vignoles*, 9; Portsmouth History Centre, Vignoles Correspondence, *passim*.

208 *a clandestine marriage*: K.H. Vignoles, *Charles Blacker Vignoles*, 13; O.J. Vignoles, *Charles Blacker Vignoles*, 65.

208 *refused to have anything more*: O.J. Vignoles, *Charles Blacker Vignoles*, 74–5;

letters of Margaret Hutton and Mary Vignoles to Charles Blacker Vignoles, 13 Dec 1816 and 16 Jan 1818, and of Charles Blacker Vignoles to Mary Vignoles, 14 and 28 Nov 1820, Portsmouth History Centre, Vignoles Papers, Letters 126, 166, 200, 202.

208 *to work as a surveyor*: Hall-Patch, 'Charles Blacker Vignoles', 234.

208 *Leybourn wrote a few months later*: O.J. Vignoles, *Charles Blacker Vignoles*, 75. But relations later deteriorated again: letters of Mary Vignoles to Charles Blacker Vignoles, 16 Apr and 17 Jun 1822, Portsmouth History Centre, Vignoles Papers, Letters 225, 227.

208 *remaining steadily implacable*: Letter of Mary Vignoles to Charles Blacker Vignoles, 1 Feb 1818, Portsmouth History Centre, Vignoles Papers, Letter 168.

208 *Isabella wrote covertly*: Letter of Isabella Hutton to Charles Blacker Vignoles, 31 Jul 1817, Portsmouth History Centre, Vignoles Papers, Letter 153/1; letters 187 and 189 (Leybourn to C.B. Vignoles, 26 Apr and 13 Sep 1819) state that Hutton later forbade Isabella to write to Vignoles.

209 *Henry Hutton sided with his father*: Hall-Patch, 'Charles Blacker Vignoles', 241; Letter of Charles Blacker Vignoles to Mary Vignoles, 3 Jun 1823, Portsmouth History Centre, Vignoles Papers, Letter 271.

12 Peace

210 *Bedford Row, London. 21 September 1822*: *Tribute of Respect*, 4.

211 *General Henry Hutton moved*: Gregory, 'Memoir', 221.

211 *Ellen was occasionally around*: Letter of Margaret Hutton to Charles Blacker Vignoles, 11 Aug 1814, Portsmouth History Centre, Vignoles Papers, Letter 72.

211 *'delighted' by the society of his children*: Gregory, 'Memoir', 221.

211 *a 'tottering scrawl'*: Charles Hutton to Catherine Hutton, 26 Jan 1819, Wellcome Collection, MS 5270 no. 35.

211 *'Cousin' Catherine kept Charles ... supplied*: Catherine Hutton, *Reminiscences*, 182–3.

211 *A portrait of Hutton in his later years*: K.H. Vignoles, *Charles Blacker Vignoles*, 4.

211 *social evenings involved poetry and music*: K.H. Vignoles, *Charles Blacker Vignoles*, 6.

211 *Hutton would make notes*: Gregory, 'Memoir', 221.

212 *a huge number of tradesmen*: Horsley, *Eighteenth-Century Newcastle*, 50.

212 *Baily on the fixing of an astronomical instrument*: Letters of Charles Hutton to Francis Baily, 17 Sep 1821, London, Senate House Library, MS 913B/3/3 (xv); and of Charles Hutton to John Bruce, 1 Jul 1814, printed in Bruce, *Memoir*, 27–9.

212 *such matters as the flow of the tide*: Charles Hutton, manuscript reply to

queries about London Bridge, 18 Jun 1819, Portsmouth History Centre, Vignoles papers, letter 186/1; London Metropolitan Archive, CLA/022/02/024: Reports to the Bridge House Estates Committee re state of London Bridge (appendices), 1821.

213 *Maskelyne . . . 'did not publish much'*: Hutton, *Dictionary*, (1815), vol. 2, pp. 22–3; John Heilbron, 'A Mathematicians' Mutiny with Morals', in Paul Horwich (ed.), *World Changes: Thomas Kuhn and the nature of science* (Cambridge, MA, 1993), 81–129 at 97.

213 *'the tears trickling down his cheeks'*: Gregory, 'Memoir', 227.

213 *Gilbert Austin and his wife and nephew*: Gilbert Austin to Charles Hutton, 13 Aug 1806, Portsmouth History Centre, Vignoles Correspondence, 1072A/ O1; British Library, Add MSS 35071 (Vignoles Journals 1830–3; 1 Jan 1830 describing a visit to two 'Children of my Grandfather's Sisters', Dr Austin and Mrs Charnley). I am most grateful to John Vignoles (personal communication) for further information.

213 *Hutton's own brothers*: Woodhorn Archive, SANT/BEQ/26/1/7/77 (notes of Thomas Wilson, 28 Mar 1822); St Bartholomew Bt, Longbenton, 20 December 1769 (burial), register accessed via www.freereg.org.uk.

213 *kept in touch with a number of friends*: letters printed in Melmore, 'Some Letters' and Bruce, *Memoir*; *A Catalogue of the . . . Library of Charles Hutton*, passim.

214 *the Bruce brothers*: letter of Charles Hutton to John Bruce, 1 Jul 1814, printed in Bruce, *Memoir*, 27–9; Wallis, 'Mathematical Tradition', 34.

214 *William Turner*: Derek Orange, 'Rational Dissent and Provincial Science: William Turner and the Newcastle Literary and Philosophical Society', in Ian Inkster and Jack Morrell, *Metropolis and Province: science in British culture, 1780–1850* (London, 1983), 202–25, esp. 211–13; 'Newcastle, Hanover Square, Unitarian Chapel', in 'Sitelines: Tyne and Wear's historic environment record', http://www.twsitelines.info/smr/7688; Wallis, 'Mathematical tradition', 47.

214 *When the Lit and Phil decided*: Letter of Charles Hutton to John Bruce, 22 Mar 1822, printed in Bruce, *Memoir*, 39–42; notice in *The Morning Post* (17 April 1822).

214 *one of the toasts*: Report in *The Newcastle Courant* (7 September 1822).

214 *a visit to the scenes of his youth*: Bruce, *Memoir*, 29, 32–3, 42–4 including letters of Charles Hutton to John Bruce, 15 Jan 1816 and 22 Mar and 20 Apr 1822.

215 *head of a learned world*: Miller, 'The Royal Society', 9; Miller, 'The Revival', 109.

216 *a generation of mathematicians*: Miller, 'Sir Joseph Banks', 289; Miller, 'The Royal Society', 108, 113–15.

216 *'petty but inextinguishable malignity'*: [Gregory], 'A Review', 245.

216 *The astronomer William Herschel*: Miller, 'The Revival', 110, citing a letter of Herschel to Babbage, 19 Dec 1820.

216 *Banks had vexed and opposed Maskelyne*: Higgitt, *Maskelyne*, 233; [Gregory], 'A Review', 246–7.

217 *Banks had organised criticism*: Olinthus Gregory, 'Vindication of the Attack on Don Rodriguez's Paper in the Philosophical Transactions', *Annals of Philosophy* 3 (1814), 282–4 (and editorial comment at 285); Miller, 'The Royal Society', 135; [Gregory], 'A Review', 249.

217 *Banks remodelled the Board of Longitude*: [Gregory], 'A Review', 246; William J. Ashworth, 'The Calculating Eye: Baily, Herschel, Babbage and the business of astronomy', *The British Journal for the History of Science* 27 (1994), 409–41 at 414.

217 *he opposed the formation*: Miller, 'Between Hostile Camps', 18; [Gregory], 'A Review', 255; Ashworth, 'The Calculating Eye', 414; Gascoigne, *Joseph Banks*, 259.

217 *Society for the Improvement of Naval Architecture*: [Gregory], 'A Review', 251–3; Miller, 'Between Hostile Camps', 7.

217 *'torturing the unseen and unknown'*: Hutton, *Dictionary*, vol. 1, p. 458.

217 *mathematician Patrick Kelly and journalist William Nicholson*: Miller, 'The Royal Society', 112; Ashworth, 'The Calculating Eye', 422; Watts, 'We Want no Authors', 405.

217 *'but an ambiguous honour'*: Miller, 'Between Hostile Camps', 12, quoting a letter of Baily to Babbage, 11 November 1820, British Library, Add. MS 37182, fol. 291.

218 *papers rejected from the* Transactions: [Gregory], 'A review', 248.

218 *Leybourn, of the Royal Military College*: See Chapter 11.

218 *Babbage held testimonials*: Charles Hutton, testimonial for Charles Babbage, 21 March 1816, British Library, Add. MS 37182, fos. 60–61; Miller, 'The Royal Society', 116–18.

218 *Hutton corresponded . . . with Waring*: letter of Charles Hutton to Francis Baily, 13 Jul 1808, London, Senate House Library, [DeM] L.4 [Waring] (copy of Edward Waring, *On the Principles of Translating Algebraic Quantities* (Cambridge, 1792); letter pasted in at end).

218 *The whole attainment*: 'Sir Joseph Banks', *New Times* (14 July 1820), 4 (with a correction to the spelling).

219 *'notoriously fond of farming'*: [Gregory], 'A Review', 166 and *passim*.

220 *William Wollaston*: Miller, 'The Royal Society', 252; Trevor I. Williams, 'Wollaston, William Hyde (1766–1828), chemist, physicist, and physiologist' in *ODNB*.

220 *A small man with a piercing glance*: David Knight, 'Davy, Sir Humphry, baronet (1778–1829), chemist and inventor' in *ODNB*; Fordyce, *A History of Coal*, 79; *The History of the British Coal Industry*, vol. 2, p. 139.

220 *he praised mathematics*: Miller, 'The Royal Society', 253.

221 *'the [highest] efforts of human intelligence'*: Royal Society, JBO/43, 150–51 (7 December 1820).

221 *He brought members of the Astronomical Society*: Miller, 'The Royal Society', 256–8; Miller, 'Between Hostile Camps', 30–31.

221 *'There is certainly no branch of Science . . .'*: Royal Society, JBO/43, 309–10 (30 November 1821).

222 *Hutton himself signed the nominations*: Royal Society, JBO/43, 177, 342 (11 January, 20 December 1821).

222 *an assistant entrusted with them*: Gregory, 'Memoir', 222; Leybourn, 'Charles Hutton', 195.

222 *an article in the French journal*: P.S. de Laplace, 'Sur la figure de la Terre, et la loi de la pesanteur à sa surface', *Connaissance des Tems* (1821), 326–31 at 330; Letter of Charles Hutton to the Marquis de Laplace, 9 April 1819, printed in the *Philosophical Magazine* ser. 1, 55 (February 1820), 81–5 and (in French) in the *Journal de Physique, de chimie, et de l'Histoire Naturelle* 90 (April 1820), 307–12; the Marquis de Laplace to Charles Hutton, 11 Sep 1820, printed (in English) in the *Philosophical Magazine*, ser. 1, 56 (November 1820), 321–2; P.S. de Laplace, 'Sur la densité moyenne de la terre', *Journal de Physique, de chimie, et de l'Histoire Naturelle* 91 (1821), 146–50, also printed in the *Connaissance des tems pour l'an 1823* (Paris, 1820), 328–31, and in English in the *Philosophical Magazine*, ser. 1, no. 56 (November 1820), 322–6.

222 *recalculated Cavendish's result*: Charles Hutton, 'On the Mean Density of the Earth', *Philosophical Transactions* 111 (1821), 276–92 (reprinted in *Philosophical Magazine* 58/279 (1821), 3–13).

223 *'no name can be mentioned'*: obituary of Hutton in the *London Courier* (February 1823).

223 *one man's patronage had been withdrawn*: [Gregory], 'A Review', 250.

223 *a plan to honour their old friend*: 'Bust of Dr. Hutton', *The Gentleman's Magazine* 91 (1821), 452; *Tribute of Respect*, 2, 3.

223 *Subscriptions poured in*: *Tribute of Respect*, 5–7.

224 *anxiety among the subscribers*: *Tribute of Respect*, 2; obituary of Charles Hutton in the *Literary Gazette* (1 February 1823), 75–6 at 76.

225 *The bust took up its station*: obituary of Charles Hutton in the *Literary Gazette* (1 February 1823), 75–6 at 75.

226 *Lord Eldon*: Letter of Lord Eldon to Henry Hutton, 3 Feb 1823, printed in Bruce, *Memoir*, 46–7.

226 *the Council slate at the Royal Society*: Royal Society, JBO/43, 522 (30 November 1822); *Literary Chronicle*, 7 December 1822, 781.

226 *Hutton never attended a meeting*: Royal Society, CMO/10.

226 *'quite the old man'*: letter of Charles Hutton to John Bruce, c. 1815, extract printed in Bruce, *Memoir*, 32–3, also 37; letter of Margaret Hutton to Charles

Blacker Vignoles, 14 Mar 1814, Portsmouth History Centre, Vignoles Papers, Letter 38; Charles Hutton to Catherine Hutton, 26 Jan 1819, Wellcome Collection, MS 5270 no. 35.

226 *the continuing vivacity*: Catherine Hutton to Charles Hutton, 4 Nov 1821, printed in Catherine Hutton, *Reminiscences*, 180.

227 *later in 1819 . . . his health improved*: Letters of Margaret Hutton to Charles Blacker Vignoles, 11 Aug 1814; of Isabella Hutton to Charles Blacker Vignoles, 24 Mar 1815; of Margaret Hutton to Charles Blacker Vignoles, 3 Aug 1816; of Mr Leybourn to Charles Blacker Vignoles, 26 Apr 1819; of Mr Leybourn to Charles Blacker Vignoles and Mary Vignoles, 13 Sep 1819, Portsmouth History Centre, Vignoles Papers, Letters 72, 80, 122, 187, 189.

227 *One visitor in the spring*: obituary of Charles Hutton in the *Literary Gazette* (1 February 1823), 75–6.

228 *the happiest year of his life*: obituary of Hutton in *The European Magazine* (June 1823), 483–7 at 484.

228 *In the winter of 1822*: Bruce, *Memoir*, 46.

228 *a letter from the Corporation of London*: Gregory, 'Memoir', 225.

228 *around four on Monday morning*: Gregory, 'Memoir', 225.

228 *He was buried the next day*: May, *Charlton*, 69.

Epilogue

229 *Dr. Hutton is gone*: John MacCulloch, *The Highlands and Western Isles of Scotland* (London, 1824), vol. 1, p. 435.

230 *condolences from both the Duke of Wellington*: Catherine Hutton, *Reminiscences*, 178.

230 *Charles Hutton Potts*: Report of resignation in *Newcastle Courant* (23 October 1863); various works of Charles Hutton Lear in for instance the National Portrait Gallery, http://www.npg.org.uk; Charles Hutton Dowling, *Metric Tables* (London, 1864).

230 *This country shall entwine*: Charlotte Caroline Richardson, Elegy on Charles Hutton, *The Ladies' Diary* (1824), 18–19 at 19.

230 *one of the most successful promoters*: Obituaries of Hutton in *The Morning Post*, 28 January 1823; and in *The Literary Chronicle* 5 (1 February 1823), 77; Humphrey Davy, *Discourse of the President . . . on the award of the Copley Medal to John Pond, Esq* (London, 1823) 59; review of Hutton, *Tracts* (1812) in *Quarterly Review* (1813), 400–18, at 418.

230 *Henry Hutton retired to Ireland*: Rogers, 'Register of the Collegiate Church of Crail', 329; obituary of Henry Hutton in *The Gentleman's Magazine* (December 1827), 561–2.

230 *a mass of papers*: Catherine Hutton, *Reminiscences*, 196–7, 202; Rogers, 'Register of the Collegiate Church of Crail', 330; H.J.C. Grierson, D. Cook

and W.M. Parker, *Letters of Sir Walter Scott* (London, 1932–7), vol. 11, pp. 295–6 and vol. 12, pp. 376–7; *Summary Catalogue of the Advocates' Manuscripts* (Edinburgh, 1971), 79, 254, 255, 599.

230 *Isabella lived quietly on*: Portsmouth History Centre, Vignoles Papers, Letters 255–84, *passim*; Catherine Hutton, *Reminiscences*, 172–97; note on the two surviving Hutton daughters, Isabella and Catherine, in *The Monthly Magazine* 26 (August 1838), 206–7; notice of Isabella's death in *The Gentleman's Magazine* (June 1839), 666.

231 *Eleanor . . . married for a third time*: Portsmouth History Centre, Vignoles Papers, family tree by K.H. Vignoles and letters 566, 863; notice of death in *The Gentleman's Magazine* (4 March 1850), 454.

231 *Vignoles, still in America*: Letter of Charles Blacker Vignoles to Mary Vignoles, 25 March 1823, Portsmouth History Centre, Vignoles Papers, Letter 261.

231 *Henry Hutton would never entirely forgive*: Hall-Patch, 'Charles Blacker Vignoles', 241; but compare the more positive view expressed (after Henry Hutton's death) in a letter of his widow to Charles Blacker Vignoles, 20 Aug 1827, Portsmouth History Centre, Vignoles Papers, Letter 365.

231 *Isabella had always had a soft spot*: Letter of Charles Blacker Vignoles to Mary Vignoles, 3 Jun 1823, Portsmouth History Centre, Vignoles Papers, Letter 271 and subsequent letters 271–84 bearing on Vignoles's relationship with Isabella, *passim*.

232 *'in memory of my Grandfather'*: British Library, Add. MS 58203, Charles Blacker Vignoles, diary: 18 November 1824.

232 *Vignoles's career took flight*: Hall-Patch, 'Charles Blacker Vignoles', 238; O.J. Vignoles, *Charles Blacker Vignoles, passim* and K.H. Vignoles, *Charles Blacker Vignoles, passim*; 'Presentation to the Royal Society', *Science*, New Series, vol. 96 (4 September 1942), 224–5 at 225.

232 *His domestic life was unhappy*: K.H. Vignoles, *Charles Blacker Vignoles*, 61; Portsmouth History Centre, Vignoles Papers, especially Letters 603, 642, 643 (quote).

233 *Vignoles's own daughter*: K.H. Vignoles, *Charles Blacker Vignoles*, 79; Portsmouth History Centre, Vignoles Papers, letters 683, 698, 699.

233 *his cousin, Henry's son*: 'Charles Hutton's Descendants', *Notes and Queries* 147 (1924), 53; two letters of Henry Hutton (Jr) bound with a copy of Bruce, *Memoir* in Newcastle Literary and Philosophical Society, N925/1; 'Charles Hutton' in *DNB* 28 (1891), 351–3 at 352.

233 *about forty-five thousand pounds*: Letter of Charles Blacker Vignoles to Mary Vignoles, 9 Jul 1823, Portsmouth History Centre, Vignoles Papers, Letter 286.

233 *His will*: The National Archives, PROB 11/1666/375, fol. 297v: Will of Charles Hutton, 27 Feb 1823.

233 *some in the family resented*: Letters of Mary Vignoles to Charles Blacker Vignoles, 15 Feb 1823 and of Charles Blacker Vignoles to Mary Vignoles, 25 Mar 1823, Portsmouth History Centre, Vignoles Papers, Letters 258, 261; Woodhorn SANT/BEQ/26/1/7/98/a: note by Thomas Wilson of 11 July 1824.

234 *a trust was set up*: scattered references in Portsmouth History Centre, Vignoles Papers, including Letters 304, 325, 336, 346, 348, 360, 361, 365, 396, 630/1, and in British Library, Add. MS 58203 (Vignoles's diary), including entries for 9 Jan, 23 Jan and 2 Mar 1824. I am grateful to John Vignoles (personal communication) for further information.

234 *he became the sole owner*: That this was Isabella's intention is suggested in a letter of Charles Blacker Vignoles to Mary Vignoles, 24 Jun 1823, Portsmouth History Centre, Vignoles Papers, Letter 281; I have not traced Isabella's will.

234 *a notoriously methodical man*: Gregory, 'Memoir', 220.

234 *There was even a diary*: Gregory, 'Memoir', 220, 221, 226–7.

235 *Hutton's novel eprouvette*: Hutton, *Tracts* (1812), vol. 3, p. 157; Jones, *Records of the RMA*, 115, xiii.

235 *His manuscript lectures on natural philosophy*: Gregory, 'Memoir', 222.

235 *His contract with the Stationers' Company*: *Records of the Stationers' Company*, Series I, Box B, folder 6, items i, iii and iv: draft contracts.

235 *sorted and labelled at least some*: Firepower MD 913/1 and 913/3, labelled for Gregory; a number of Hutton's surviving letters are labelled with the names of their senders and dates in Hutton's hand.

235 *several more scientific items*: *Catalogue of a Miscellaneous Collection of Books: being the valuable and scientific library of the late Dr. Olinthus Gregory . . . which will be sold by auction by Messrs. Southgate and Son . . . on Thursday, March the 17th, 1842 and following day* [London, 1842], 9, 24 (these items are now in Trinity College Cambridge, MS R.1.59), 25; London, Senate House Library MS 235 (Hutton's translation from Tartaglia), endorsement on fol. 1r; MS 913B/3/1 (xiv, xv), letters of John Playfair to Charles Hutton, 12 Dec 1782, of Sir John Leslie to Charles Hutton, 14 Oct 1795, and of Charles Hutton to Francis Baily, 1821: endorsements.

236 *Gregory had the notion of working*: Trinity College, Cambridge, MS R.1.59, fos. 179–217.

236 *that Hutton's diary would appear in print*: Gregory, 'Memoir', 220; Bruce, *Memoir*, 1; Mackenzie, *Historical Account of Newcastle*, 560.

236 *A few items turned up*: *A Catalogue of the Scientific Manuscripts in the Possession of J.O. Halliwell* [n.p. n.d.], 6 (historical papers); [Sotheby], *Catalogue of a Collection of Scientific and Historical Manuscripts* [a portion of the collection of James Orchard Halliwell] [London, 1840], 12 (letters and papers on the history of logarithms); *Original Letters, Manuscripts, and State Papers, Collected by William Upcott* ([London], 1836), 45 (letters), 50

(assignments of copyrights); also *A Bibliography of the Works of Sir Isaac Newton: together with a list of books illustrating his works with notes by George J. Gray* (Cambridge, 1907), 26 (Hutton's manuscript library catalogue).

236 *'arranging papers'*: British Library, Add. MS 35071, 4, 12 and 24 January; also Add. MS 58203 (Vignoles's diary), 2 and 4 February 1824: 'very particular business for my Aunt'; letter of Charles Blacker Vignoles to the Marquis of Northampton, 25 Mar 1841, Portsmouth History Centre, Vignoles Papers, Letter 751 (quote); letter of Charles Blacker Vignoles to Francis Galton, 17 Nov 1865, London, University College archives, MS Galton 2/4/1/2/9 (quote).

237 *part of Newcastle University*: Wallis, 'Mathematical Tradition', 34.

237 *His own houses on Woolwich Common*: Saint and Guillery, 'Woolwich Common', 25–6; Jones, *Records of the RMA*, 111.

237 *'Military tactics have been much benefited'*: obituary of Hutton in *London Magazine* 7 (March 1823), 368.

238 *eminently contributed to awaken*: Davy, *Discourse of the President*, 60.

238 *utmost importance to the British nation*: Review of Hutton, *Dictionary* in *Critical Review* 18 (November 1796), 302–5 at 304.

238 *Davy's reforms at the Society stalled*: Knight, 'Davy'.

238 *'science in England is not a profession'*: Charles Babbage, *The Exposition of 1851* (London, 1851), 189.

239 *a 'competent knowledge of the first book of Euclid'*: Warwick, *Masters of Theory*, 64.

239 *Fellow of the Royal Astronomical Society*: The printed lists of fellows from this period do not mention Hutton; one is dated in February 1822, before his election, and one in February 1824, after his death: *Memoirs of the Royal Astronomical Society* 1/1, 225ff; 1/2, 523ff. See also www.ras.org.uk (election on 10 May 1822); *Memoirs* 1/1 (1822), 219, 222 (notice of gifts); *Memoirs* 1/2 (1825), 498 (mention of death), 515 (notice of gift).

239 *dedication of a book to him*: for instance Henry Clarke, *A Dissertation on the Summation of Infinite Converging Series with Algebraic Divisors* (London, 1779); William Marrat, *Introduction to the Theory and Practice of Mechanics* (Boston, 1810); Charlotte Caroline Richardson, *Harvest; a poem: with other poetical pieces* (London, 1818).

239 *values of the Royal Society*: Rebekah Higgitt, 'Why I Don't FRS my Tail: Augustus De Morgan and the Royal Society', *Notes and Records of the Royal Society* 60 (2006), 253–9.

240 *'arch of equilibration'*: Alberto Cecchi, 'The "Arch of Equilibration" of Charles Hutton (1772)', *Meccanica* 45 (2010), 829–33, esp. 832.

240 *the torsion balance experiments of Cavendish*: Jungnickel and McCormmach, *Cavendish*, 449–50; Francis Baily, 'Experiments with the Torsion Rod for Determining the Mean Density of the Earth', *Memoirs of the Royal*

Astronomical Society 14 (1843), 1–130 and i–cclviii, at 92–6; J.H. Poynting, *The Mean Density of the Earth* (London, 1894), 18; Smallwood, 'Maskelyne's 1774 Schiehallion Experiment', 15, 28–30; John R. Smallwood, 'John Playfair on Schiehallion, 1801–1811', in C.L.E. Lewis and S.J. Knell, *Making of the Geological Society of London* (London, 2009), 279–97 at 293; Howse, *Maskelyne*, 141; G.S. Leadstone, 'Maskelyne's Schiehallion Experiment of 1774', *Physics Education* 9 (1974), 452–8.

241 *'Unwin's formula'*: Robins Fleming, *Six Monographs on Wind Stresses; wind pressure factors, specification* (New York, 1915), 6. A discussion of the later use and extension of Hutton's ballistics work is in David Aubin, 'Ballistics, Fluid Mechanics, and Air Resistance at Gâvre, 1829–1915: doctrine, virtues, and the scientific method in a military context', *Archive for History of Exact Sciences* 71 (2017), 509–42.

241 *'improving and simplifying'*: *Tribute of Respect*, 2; cf. Margaret Baron, 'Charles Hutton' in *DSB*; D.E. Smith, *History of Mathematics* (Boston, 1923–5) vol. 2, p. 458.

241 *'the most useful and important'*: 'Memoir of the late Dr. Hutton', *The Gentleman's Magazine* (March 1823), 228–32 at 229.

241 *four volumes of local Newcastle history*: Sykes, *Local Records* (1824); Mackenzie, *Historical Account of Newcastle*; Boyle, *Vestiges of Old Newcastle*; Welford, *Men of Mark*.

242 *a few reminiscences in manuscript*: Woodhorn, SANT/BEQ/26/1/7/77. Thomas Wilson, notes on Mr. Kirkley's reminiscences of Charles Hutton, 28 Mar 1822; SANT/BEQ/26/1/7/98/a. Thomas Wilson, notes including information about Charles Hutton. 11 Jul 1824; SANT/BEQ/26/1/8/584. Letter, S. Barrass to Thomas Wilson including George Parkin's reminiscences of Charles Hutton. 1825.

242 *Henry quashed the biography*: H.P. Chanter, 'Charles Hutton's Descendants', *Notes and Queries* 146 (June 1924), 471, citing an unnamed 'old handbook on Newcastle'.

242 *'Collections relative to Charles Hutton'*: Birmingham Archives, Heritage and Photography Service, MS 3597/103/3 (title page only).

243 *The Course also went to America*: Rickey and Shell-Gellasch, 'Mathematics Education at West Point'.

243 *Translations were arranged*: All of these can be found in the catalogues of large libraries today; some of them are mentioned in J.F. Blumhardt, *Catalogue of the Library of the India Office*, vol. 2, part 5: *Marathi and Gujarati Books* (London, 1908), pp. 89, 91, 245–6; or Charles Ambrose Storey, *Persian Literature: a bio-bibliographical survey*, vol. 2, issue 1 (1927), 19.

243 *A list of authors*: A.G. Howson, 'Charles Hutton', in *A History of Mathematics Education in England* (Cambridge, 1982), 59–74 at 69.

243 *John Henry Newman*: Ian Turnbull Ker and Thomas Gornall (eds), *The*

Letters and Diaries of John Henry Newman, vol. 7 (Oxford, 1978), p. 17, n. 1; *Collected Works of Marx and Engels* vol. 33 (New York, 1975), 522, 524 (n. 223).

243 *'an interesting and learned history'*: 'Charles Hutton' in *DNB* vol. 28 (1891), 351–3 at 352.

Image Credits

Figure 1, page 4
J.R. Boyle, *Vestiges of Old Newcastle and Gateshead* (Newcastle, 1890), p. 150.
Photograph by the author from a copy in the author's collection.

Figure 2, page 15
The Condition and Treatment of the Children employed in the Mines and Collieries of the United Kingdom (London, 1842), frontispiece.
Wellcome Collection. CC BY.

Figure 3, page 20
George Fisher, *The Instructor* (London, 1760), frontispiece.
World History Archive/Alamy Stock Photo.

Figure 4, page 25
J. Draper, *Young Students' Pocket Companion* (Whitehaven, 1772), frontispiece.
The Bodleian Libraries, The University of Oxford: (OC) 181. G. 122.

Figure 5, page 30–1
Nathaniel and Samuel Buck, *The South-East Prospect of Newcastle Upon Tyne* (1745).
© Image; Crown Copyright: UK Government Art Collection March 2014.

Figure 6, page 44–5
Charles Hutton, *A Plan of Newcastle-upon-Tyne* (Newcastle, 1772), detail.
The Bodleian Libraries, The University of Oxford: Gough Maps Northumberland 20.

Figure 7, page 54
Paul Sandby, *The Royal Military Academy at Woolwich*, c. 1770.
Royal Collection Trust/© Her Majesty Queen Elizabeth II 2018.

Figure 8, page 56
Thomas Seccombe, *Gentlemen Cadets*, 1783.
Royal Collection Trust/© Her Majesty Queen Elizabeth II 2018.

Figure 9, page 72
Edward Scriven, *The Gallery of Portraits with Memoirs* (1833–7), vol. 6, plate between pp. 20–21.
The Print Collector/Alamy Stock Photo.

Figure 10, page 89
Charles Hutton, 'The Force of Fired Gun-Powder, and the Initial Velocities of Cannon Balls, Determined by Experiments', *Philosophical Transactions* 68 (1778), accompanying plate.
Photograph by the author from a copy in the author's collection.

Figure 11, page 97
Joshua Reynolds, *Sir Joseph Banks*, 1773.
The Picture Art Collection / Alamy Stock Photo.

Figure 12, page 115
Engraving of Charles Hutton by Charles Knight after Mary Byrne.
Wellcome Collection. CC BY.

Figure 13, page 147
Philip Reinagle, *Isabella and Camilla Hutton*, 1788.
Photograph by Jim Campbell. Reproduced by kind permission of the Vignoles family.

Figure 14, page 159
Portsmouth Record Office, Vignoles papers 1072A/App. 10, fol. 1.
Reproduced by kind permission of Portsmouth Libraries & Archives Service, Portsmouth City Council, All rights reserved.

Figure 15, page 176
Engraving of Olinthus Gregory by Henry Robinson after Richard Evans.
Science History Images/Alamy Stock Photo.

Figure 16, page 190
Engraving of Charles Hutton after H. Ashby.
Science History Images/Alamy Stock Photo.

Figure 17, page 216
Thomas Phillips, *Sir Joseph Banks*, 1810.
Granger Historical Picture Archive/Alamy Stock Photo.

Figure 18, page 225
Engraving of Hutton's bust after Sebastian Gahagan, 1822.
Chronicle/Alamy Stock Photo.

Figure 19, page 227
Andrew Morton, Charles Hutton, c. 1822.
Photograph reproduced by permission of the Newcastle Literary
and Philosophical Society.

Index